# The Use of Online Collaboration Tools for Employee Volunteering: A Case Study of IBM's CSC Program

# RIVER PUBLISHERS SERIES IN INNOVATION AND CHANGE IN EDUCATION - CROSS-CULTURAL PERSPECTIVE
Volume 10

*Series Editor*
**XIANGYUN DU**
*Aalborg University*
*Denmark*

**Editorial Board**

- **Alex Stojcevski**, Faculty of Engineering, Deakin University, Australia
- **Baocun Liu**, Faculty of Education, Beijing Normal University, China
- **Baozhi Sun**, North China Medical Education and Development Center, China
- **BinglinZhong**, Chinese Association of Education Research, China
- **Bo Qu**, Research center for Medical Education, China Medical Education, China
- **Danping Wang**, The Department of General Education, Technological and Higher Education Institute of Hong Kong
- **Fred Dervin**, Department of Teacher Education, Helsinki University, Finland
- **Kai Yu**, Faculty of Education, Beijing Normal University, China
- **Jiannong Shi**, Institute of Psychology, China Academy of Sciences, China
- **Juny Montoya**, Faculty of Education, University of ANDES, Colombia
- **Mads Jakob Kirkebæk**, Department of Learning and Philosophy, Aalborg University, Denmark
- **Tomas Benz**, Hochschule Heilbronn, Germany

Nowadays, educational institutions are being challenged when professional competences and expertise become progressively more complex. This is mainly because problems are more technology-bounded, unstable and ill-defined with the involvement of various integrated issues. To solve these problems, it requires interdisciplinary knowledge, collaboration skills, innovative thinking among other competences. In order to facilitate students with the competences expected in professions, educational institutions worldwide are implementing innovations and changes in many aspects.

This book series includes a list of research projects that document innovation and change in education. The topics range from organizational change, curriculum design and innovation, pedagogy development, to the role of teaching staff in the change process, students' performance in the aspects of not only academic scores, but also learning processes and skills development such as problem solving creativity, communication, and quality issues, among others. An inter- or cross-cultural perspective is studied in this book series that includes three layers. First, research contexts in these books include different countries/regions with various educational traditions, systems and societal backgrounds in a global context. Second, the impact of professional and institutional cultures such as language, engineering, medicine and health, and teachers' education are also taken into consideration in these research projects. Thirdly, individual beliefs, perceptions, identity development and skills development in the learning processes, and inter-personal interaction and communication within the cultural contexts in the first two layers.

We strongly encourage you as an expert within this field to contribute with your research and make an international awareness of this scientific subject.

For a list of other books in this series, www.riverpublishers.com
http://www.riverpublishers.com/series.php?msg=Innovation_and_Change_in_Education_-_Cross-cultural_Perspective

# The Use of Online Collaboration Tools for Employee Volunteering: A Case Study of IBM's CSC Program

Ayse Kok

Oxford University

*Published, sold and distributed by:*
River Publishers
Niels Jernes Vej 10
9220 Aalborg Ø
Denmark

River Publishers
Lange Geer 44
2611 PW Delft
The Netherlands

Tel.: +45369953197
www.riverpublishers.com

ISBN: 978-87-93379-17-6 (Hardback)
      978-87-93379-16-9 (Ebook)

©2016 River Publishers

All rights reserved. No part of this publication may be reproduced, stored in a retrieval system, or transmitted in any form or by any means, mechanical, photocopying, recording or otherwise, without prior written permission of the publishers.

# Contents

Foreword ... ix

Preface ... xi

Acknowledgments ... xv

List of Figures ... xvii

List of Tables ... xix

List of Abbreviations ... xxi

Abstract ... xxiii

1 Introduction ... 1
  1.1 Asking the Why Question ... 2
  1.2 Asking the How Question ... 5
  1.3 Asking the What Question ... 6

2 Some Theories about Online Collaboration ... 9
  2.1 Workplace Learning ... 10
  2.2 The Social Aspects of Workplace Learning ... 12
    2.2.1 Summary for WPL ... 14
  2.3 Exploring CSCL in Depth ... 15
    2.3.1 A Crucial Definition: Cooperation versus Collaboration ... 17
      2.3.1.1 Collaborative learning ... 18
    2.3.2 Collaborative Learning Tools ... 21
    2.3.3 Groupware ... 23
    2.3.4 New Generation of Collaborative Learning Tools ... 25
    2.3.5 Summary for CSCL Theories ... 27
  2.4 Conceptual Framework ... 28
  2.5 Summary ... 30
  2.6 Concluding Remarks ... 31

## 3 Researching the Online Experiences of IBM Employees  33
3.1 Research Questions . . . . . . . . . . . . . . . . . . . . . . . . . 33
3.2 The Origins of the Methods Used for Exploring the IBM CSC Experience . . . . . . . . . . . . . . . . . . . . . . . . . . . . . . . 33
    3.2.1 Participatory Design . . . . . . . . . . . . . . . . . . . . . 34
    3.2.2 Participatory Research . . . . . . . . . . . . . . . . . . . . 34
3.3 Definition of Participatory Research in the Context of CSC Program . . . . . . . . . . . . . . . . . . . . . . . . . . . . 35
    3.3.1 Overview of the Stages of Participation . . . . . . . . . . . 36
        3.3.1.1 Involvement of project partners . . . . . . . . . . 37
        3.3.1.2 Involving senior managers of the CSC program . . . . . . . . . . . . . . . . . . . . . . . 38
3.4 Data Collection . . . . . . . . . . . . . . . . . . . . . . . . . . . . 38
    3.4.1 Description of Stage One . . . . . . . . . . . . . . . . . . . 41
    3.4.2 Description of Stage Two . . . . . . . . . . . . . . . . . . . 44
    3.4.3 Description of Stage Three . . . . . . . . . . . . . . . . . . 48
        3.4.3.1 Transcript validation . . . . . . . . . . . . . . . . 49
        3.4.3.2 Design of CSC collaboration platform . . . . . . 50
3.5 Data Analysis . . . . . . . . . . . . . . . . . . . . . . . . . . . . . 50
3.6 The Challenges of Using Participatory Approaches . . . . . . . . . 51
3.7 Ethical Considerations . . . . . . . . . . . . . . . . . . . . . . . . 54
    3.7.1 Gaining Access to Participants . . . . . . . . . . . . . . . . 54
    3.7.2 Anonymity and Confidentiality . . . . . . . . . . . . . . . . 55

## 4 Results  57
4.1 Contextual Background Information about the Tools Used by CSC Participants . . . . . . . . . . . . . . . . . . . . . . . . . . 57
4.2 Participants' Use of the Online Collaboration Tools . . . . . . . . . 58
4.3 From Online Collaboration toward Distributed Cognition . . . . . . 60
    4.3.1 Participant Approaches . . . . . . . . . . . . . . . . . . . . 61
4.4 Decisions for Using Online Collaboration Tools . . . . . . . . . . . 62
    4.4.1 Personal Factors that Influence Decisions . . . . . . . . . . 64
    4.4.2 Affordances of Technologies that Influence Decisions for Collaboration . . . . . . . . . . . . . . . . . . . . . . . 66
    4.4.3 Properties of Technologies that Influence Decisions . . . . . 67
    4.4.4 Decision for not Using Digital Tools . . . . . . . . . . . . . 69
    4.4.5 Preference for Using Certain Digital Tools (Driven by Desire) . . . . . . . . . . . . . . . . . . . . . . . 70
4.5 Evaluation of Support . . . . . . . . . . . . . . . . . . . . . . . . . 71
    4.5.1 Reasons for not Receiving Any Technology Support . . . . 72
        4.5.1.1 Hit and miss . . . . . . . . . . . . . . . . . . . . 72
    4.5.2 Evaluation of Informal Sources of Support . . . . . . . . . 73
4.6 Understanding Perceptions and Beliefs of the CSC Participants . . . 74
    4.6.1 Perceptions of Technologies and Their Use . . . . . . . . . 75
    4.6.2 Beliefs about the Value of Online Collaboration Tools . . . 76

|  |  | 4.6.2.1 | Technology provides ground for easier collaboration | 76 |
|  |  | 4.6.2.2 | Positive thoughts about how online collaboration tools are used | 77 |
|  | 4.6.3 | Mixed Feelings about the Value of Online Collaboration Tools | | 81 |
|  | 4.6.4 | Defining Appropriation and Influential Factors for Intersubjectivity and the Practice | | 82 |
|  |  | 4.6.4.1 | Appropriation with regard to digital tools and activity types | 83 |
|  |  | 4.6.4.2 | Types of learning | 84 |
|  |  | 4.6.4.3 | The shifting volunteering practice | 86 |

**5 Some Initial Reflections** — **89**
 5.1 The Generalizability of the Results — 97
 5.2 Evaluation of the Methodology — 99

**6 Lessons Learnt** — **101**

**7 Concluding Remarks** — **105**

**Appendix** — **107**

**Glossary** — **175**

**References** — **181**

**Index** — **199**

**About the Author** — **201**

# Foreword

E-collaboration is about sharing information within and between organizations for the purposes of planning, coordinating, decision making, process integration, and improving efficiency and effectiveness. These technologies include Web-based chat tools, Web-based asynchronous conferencing tools, e-mail, LISTSERVS, collaborative writing tools, group decision support systems, teleconferencing suites, social networking platforms, and Web 2.0 technologies.

Organizations exchange information through people, process and technology, and increasingly rely on e-collaboration technologies to make that happen. The connectivity being made possible by the nature of the digital space that offers much potential for becoming engaged in a joint activity. Regardless of the benefit of these technologies offered to employees, be it socializing, networking, or support, much remains to be explored about these places in terms of learning and development.

This book fills an important gap by focusing on all forms of e-collaboration whether it be through e-mail, online communities of practices, social networks, Web 2.0 tools, or virtual teams in organizational settings.

Topics covered include:

- Theory that informs practice—emerging models and understanding from academia;
- Research—new understandings of learning, collaborative sense-making, and learning preferences;
- The Practitioner view—real examples from around the world of groundbreaking developments in e-collaboration that are transforming education, adult learning and corporate training;
- Guidance for designers and producers—pedagogical advice and skills for a range of people who may have had little exposure to the body of knowledge surrounding collaboration design;
- Looking to the future—what to expect in the next 5–10 years and how to prepare to take full advantage of the opportunities that an increasingly connected society will provide for computer-supported collaboration.

This book has been written with the intent of providing practical advice from academics, researchers, practitioners, and designers who are currently engaged in defining, creating, and delivering the increasingly important world of electronic collaboration. By leaving aside trends in technology, this book instead focuses on the articulation and development of the computer-supported collaboration theories that underpin the use of technology.

This book addresses key gaps in the available literature including the inequality of access to technologically enabled learning and cutting-edge design issues and pedagogies that will take us into the next decade of e-collaboration and future Web 3.0+ approaches.

The challenges in e-collaboration are both difficult and interesting. People are working on them with enthusiasm, tenacity, and dedication to develop new methods of analysis and provide new solutions to keep up with the ever-changing threats. In this new age of global interconnectivity and interdependence, it is necessary to provide practitioners, both professionals and students, with state-of-the art knowledge on the frontiers in e-collaboration. This book is a good step in that direction.

The Prophet Mohammad (s.a.w) said 'Seek knowledge even unto China'. To gain the academic knowledge on e-collaboration, we tried to collaborate with scholars across the world, whose efforts the editors of this book very much appreciate.

# Preface

Today's workplace settings are in constant need of recurrent learning processes interwoven with daily tasks on digital spaces. However, these digital spaces are not devoid of any issues and hence suggest the need for employees to be conscious of the emerging issues.

Effective collaboration will increasingly be strategic differentiators for organizations of all types in the 21st century. Information and communication technologies have a critical role to play in helping organizations to achieve these goals. By publishing new theoretical and practical research findings, as well as providing a forum for broader discussion, this book contributes to the understanding and advancement of this important domain of electronic collaboration.

This book addresses the design and implementation of e-collaboration technologies, assesses the behavioral, cultural, and social impacts of these technologies on individuals and groups, and presents theoretical considerations on links between the use of e-collaboration technologies and behavioral, social, or cultural patterns.

Research studies undertaken by Future Lab in 2007 have suggested that when people make a choice or decision not to use technology, even though access is available to them, then they are making an empowered choice. Above and beyond having the necessary access to online tools, online collaboration, therefore, is predicated on the ability to make an informed choice when and when not to make use of these tools. Online collaboration is not, therefore, simply a matter of ensuring that all individuals make use of these tools throughout their day-to-day lives, but a matter of ensuring that all individuals are able to make what could be referred to as "smart" use of technology, i.e., using it as and when appropriate. In this sense, Future Lab concluded that not making use of an online tool can be a positive outcome for some people in some situations, provided that the individual is exercising an empowered "digital choice" not to do so. The results of various studies in this book offer examples of empowered choices being made by uses; for example, choosing to use the Web 2.0 tools such as the wiki or blog because making use of them increased information-sharing, which supported the participants' progress with their projects in an organizational or academic setting. However, there are also times when participants are choosing not to use these tools because they have a preference for the more conventional methods such as face-to-face discussions or brainstorming. Various data also suggest areas that would be worthy of further exploration in terms of understanding whether or not the decisions made are empowered ones or not. A good example of this would be to provide meaningful and relevant information about how much "time" might be saved in the long run in terms of efficiency and improved collaboration outcomes. The results, therefore, build on existing theories

and discourses regarding the use of online collaboration tools, but also challenge us to expand our understanding and application of these theories with regard to the following areas:
- Swapping and changing from a range of online collaboration tools;
- Being well-informed about the strengths and weaknesses of particular online collaboration tools in relation to usability and impact on learning;
- Developing a range of sophisticated and tailored strategies for using online collaboration tools to support their learning;
- Being extremely familiar with technology; and
- Being aware of what help and support is available.

## The Challenges

This book does not compare the impact of one collaboration medium over another and does not include the implications of mature use of collaboration technologies, which have been widely discussed elsewhere. Mature users of collaboration technology are more likely to report business benefits and staff benefits of their implementation and are also more likely both to align learning and collaboration with business goals and to measure success (Bersin Associates, 2008; Towards Maturity, 2009). Mature users are also more likely to blend collaboration and learning technologies within other approaches to develop business solutions.

This book does not attempt to isolate the specific role of technology in the blend, the maturity of the user, nor the process of business alignment, all of which influence results. Caution should be used in applying the results in this review; increasing the use of technologies in collaborative learning will not automatically achieve success or efficiency savings.

## Searching for a Solution

The use of learning technologies in the workplace is on the increase. Those who are investing in learning technologies expect more from their investment. But is there clear and concrete evidence to illustrate that technology-supported collaboration and learning in the workplace actually delivers the type of bottom-line business benefits that organisations are looking for? This is the question that we investigate within this book.

The authors set out to identify and review a range of literature (academic, research, case studies, online, and print) to look for examples of both workplace and academic implementation of collaboration technologies that have tangible results. Articles, reports and case studies have been investigated using the following parameters:
- Employers—The review focused on the business users of collaboration technologies and aimed to include small and large companies encompassing the private, public and third sectors.
- Technology—The definition of technology-supported collaboration/e-collaboration used in this review includes the application of collaboration technologies

across the learning process from assessment of organizational and individual need to delivery of learning, learner support, management and administration, and formal and informal learning.

To identify information resources that address all of these three parameters, we approached an extensive network of global providers, experts (academic and non-academic), and employer membership groups in addition to conducting traditional literature research.

# Acknowledgments

I wish to personally thank all of the participants for their contributions to my inspiration and knowledge and other help in creating this book. Thank you to all who without their contributions and support this book would not have been written. I also acknowledge the valuable contributions of the reviewers regarding the improvement of quality, coherence, and content presentation of chapters. Most of the authors also served as referees; we highly appreciate their double task.

A special thank you note should be provided for former IBM Senior CSC Program Manager Kevin Thomson, putting me into contact with relevant IBM employees and other gatekeepers with regard to this book. This collaborative writing process could not have been possible without his support.

I beg forgiveness of all those who have been with us over the course of the past months while writing this book and whose names we have failed to mention.

Last and not least: My dear parents, no words of gratitude would be enough to express my gratitude for them.

# List of Figures

| | | |
|---|---|---|
| **Figure 2.1** | Graphical representation of individual and social theories of learning (Stahl, 2001). | 16 |
| **Figure 2.2** | The differences between **Web 1.0** and **Web 2.0** (O'Reilley, 2005). | 26 |
| **Figure 3.1** | Fajerman and Treseder (2000)'s model for levels of participant involvement. | 37 |
| **Figure 3.2** | Questions presented to Phase One participants via e-mail. | 43 |
| **Figure 3.3** | Example of the quality and quantity of feedback given by a participant in the pilot study. | 44 |
| **Figure 4.1** | Preferred usage of tools. | 58 |
| **Figure 4.2** | Q8: Do you think you are able to obtain sufficient information from all other CSC volunteers? | 59 |
| **Figure 4.3** | Q9: What is/are the prohibition/s in case you think you are not able to provide, share and acquire best practices, knowledge and leverage the experience of CSC volunteers? | 60 |
| **Figure 4.4** | Overview of Lotus Notes collaboration space. | 74 |
| **Figure 5.1** | Coding sub-themes identified for choices. | 94 |
| **Figure 5.2** | Coding sub-themes identified for experiences. | 95 |
| **Figure 5.3** | Coding of the sub-themes specified for the evaluation of support. | 96 |
| **Figure 5.4** | Coding sub-themes identified for feelings and beliefs. | 96 |
| **Figure 5.5** | Model for online collaboration. | 98 |
| **Figure 5.6** | Overview of the new CSC platform. | 100 |
| **Figure 5.7** | Menu items of the new CSC platform. | 100 |

# List of Tables

| | | |
|---|---|---|
| Table 2.1 | Differences between the traditional and collaborative **e-learning** model (Mandl and Krause, 2001) | 22 |
| Table 2.2 | Main elements of the conceptual framework | 29 |
| Table 3.1 | Breakdown of data collected | 39 |
| Table 3.2 | Frequency of responses (tick marks) to each proposed research question | 45 |
| Table 3.3 | Alignment between the research questions and interview questions | 47 |
| Table 3.4 | Comparison of the frameworks for preliminary and final interview coding | 49 |
| Table 3.5 | Overview of the alignment of the suggested interview coding categories with research questions and interview questions | 52 |
| Table 4.1 | Use of collaborative tools used among CSC participants | 62 |
| Table 4.2 | Overview of the types of approaches used by CSC volunteers | 63 |
| Table 5.1 | Mapping between the first research questions and the findings | 90 |
| Table 5.2 | Mapping between the second research question and the findings | 92 |
| Table 5.3 | Questions for exploring online collaboration for 'Technology-enhanced Volunteering' | 99 |

# List of Abbreviations

| | |
|---|---|
| AM | The Acquisition Metaphor |
| CAI | Computer-Aided Instruction |
| CAL | Computer-Aided Learning |
| CMC | Computer-Mediated Communication |
| CSC | Corporate Service Corps |
| CSCL | Computer-Supported Collaborative Learning |
| CSCW | Computer-Supported Collaborative Work |
| IBM | International Business Machine |
| PC | Personal Computer |
| PM | The Participation-Metaphor |
| TAM | Technology Acceptance Model |
| UNDP | United Nations Development Program |
| USB | Universal Serial Bus |
| VLE | Virtual Learning Environment |
| VPN | Virtual Private Network |
| WPL | Workplace Learning Literature |
| WWW | World Wide Web |

# Abstract

This book conveys my experience about the use of online collaboration tools in supporting knowledge workers for the practice of employee volunteering. Having gained several years of experience in the field of ICT (Information Communication Technologies) I was always curious about the use of digital technologies for the common good that would benefit others rather than merely sharing information. By online collaboration tools I refer to the web-based technologies such as popular Web 2.0 tools like blogs or wikis and traditional online tools such as instant messenger, discussion forums, online chats and e-mail used by several individuals with the aim of achieving a common goal.

The employee volunteering program-called Corporate Service Corps (CSC) – is an employee volunteering program in which the IBM employees tackle the economic and societal issues of the less developed countries they have been sent to while getting involved in project-based learning activities. This project provided me with the opportunity to provide an insight into how online engagement enabled the continuation of non-formal workplace learning practices such as volunteering and opened up possibilities for new ways to contribute to the learning process of employees. When it comes to online communities there is a mixture of entanglements, partnerships, negotiations and resistances between these tools and human actors. This book explores how online communities are created by employee volunteers and also provides an understanding of non-formal learning practices within such fluid settings; important issues for organizations interested in non-formal learning practices of their employees are also raised.

Today's workplace settings are in constant need of recurrent learning processes interwoven with daily tasks on digital spaces. However, these digital spaces are not devoid of any issues and hence suggest the need for employees to be conscious of the emerging issues. The results from the case study are analysed by using participatory design methods in order to contribute to the understanding of the use of technology as both a single and collective experience.

This research identified the specific benefits of online collaboration tools, and explored how their usage has been appropriated by employee volunteers for their practice of volunteering and how they influenced the process of their meaning-making. By doing so, it raised an awareness of the digital tools that provide collections of traits through which individuals can get involved in non-formal learning practices by having digital interactions with others.

# 1

# Introduction

Much has been said about how our lives are changed by technology including our work lives. There are more than 2 billion active **Facebook** users who spend more than 5 billion minutes on a daily basis (Facebook, 2014). Apparently, these digital spaces have become an important aspect of the lives of many people. So, it has become common sense for most employees to browse through various digital spaces such as Facebook, Twitter, and LinkedIn.

Workplace learning is mostly a social process. According to Gibb and Fenwick (2008), workplace learning is "manifested within a joint activity emerging from the dynamic relationships among individuals and groups" (p. 4). This inherent nature of connectivity of the digital tools offer much potential for knowledge workers becoming engaged in a joint activity. Regardless of the benefit of online communities offered to knowledge workers, be it socializing, networking or support, much remains to be explored about these places in terms of means of **non-formal learning**.

This study provides an insight into how online engagement enabled the continuation of non-formal workplace learning practices such as volunteering and opened up possibilities for new ways to contribute to the learning process of knowledge workers. Today's workplace settings are in constant need of recurrent learning processes interwoven with daily tasks on digital spaces. However, these digital spaces are not devoid of any issues and hence suggest the need for employees to be conscious of the emerging issues. As every knowledge-intensive entity needs to support their employees' development in non-classroom and non-instructional type of learning the crucial aspect of digital applications in terms of contributing to related processes of knowledge creation by means of **collaboration** needs an emphasis. While doing this I reflect upon the strategies adopted in alignment with the umbrella term of "**Web 2.0**".

Book explores how online communities are created by employee volunteers and also provides an understanding of non-formal learning practices within such fluid settings; important issues for organizations interested in non-formal learning practices of their employees are also raised.

This study includes several project groups making use of various collaboration tools which these group members 'appropriated' (Rogoff, 1990) to contribute to their own volunteering practice. The study conveys a context-driven collaboration model with a focus on learning through collaboration throughout a volunteering program. This volunteering program matches communities' needs in the developing world to IBM employees' learning processes in a collaborative and integrated manner. This volunteering model involves a decentralized, employee-generated learning process that is driven by collaboration with colleagues, online resources, and experts within the organizational setting in IBM. I identify the affordances of various digital tools from the perspective of employee volunteering, and how these affordances can be leveraged to support employee choice and autonomy. The volunteers made a decision for using these online collaboration tools on their own without being under the influence of any institution, and they utilized these tools completely based on their own needs and ideas. In addition to being a generic space for sharing documents, the digital environment serves as a joint place populated and created by the volunteers to navigate through information, find personal routes and pathways. This set of tools provides contextual information in a seamless manner based on the learning needs of the IBM employees. My inquiry in this book related to different volunteering cases that deal with the changing patterns of use. I delve into the collaborative processes facilitated by the use of digital tools within their volunteering context, in other words, whether and how volunteers were supported by the content conveyed to them via means of relevant digital assets and tools. The volunteering setting embeds aspects of both virtual and physical parts of workplace learning.

The following sections provide the rationale, aim and contextual background for this research including the relevant literature about both **workplace learning** and **digital learning,** as well as their underlying theoretical perspectives.

## 1.1 Asking the Why Question

Before proceeding with any study, we should also reflect on why we want to delve into that particular area and how we aim to do so. In other words, these big questions of why, what and how need to be tackled first before conveying our perspective on a particular topic.

While some scholars argue that Web 2.0 relates to the collectivity of innovative web applications (Alexander, 2006) others argue that rather than a mere group of web services, Web 2.0 entails a set of new practices (Dohn, 2010). My rationale for conducting this study is in alignment with Ryberg's (2008) statement that the adoption of new technologies includes "a synergy among all of pedagogical, organizational and technological understandings of theory and practice" (p. 664). In a similar vein, Anderson (2007, p. 4) asserts that due to the opportunity provided by the Web for establishing social connections different ways of knowing emerge as users can contribute as much as they prefer to. Edwards and Usher (2008, p. 120) describe this new type of knowledge as more "complex and possessing less hierarchy"). Due to the emergence of this new type of knowing definition of learning and collective knowledge construction also change (Haythornthwaite, 2008). As Young (2006, p. 257) stated, the "sociability" around Web 2.0, rather than the characteristics of the application itself contribute to the creation of these new ways of knowing.

The emphasis on collective knowledge construction leads to the premise that online community is an important characteristic of a quality workplace learning experience. Due to the rhetoric created by commercial agendas who feel enthusiastic about the technology itself technologies are often studied in an overly deterministic way. A plethora of studies have been conducted with the objective of conducting a research about the interrelation between the digital tools and individuals which lead to new ways of learning within the context of organizational settings (Brown and Duguid, 1991, 1998; Cohen and Levinthal, 1990; Cook and Brown, 1999); yet none of these studies has focused on the distributed volunteering practice of knowledge workers. This study aims to reveal a more complete picture of online interactions as individuals attain a greater level of expertise and co-construct knowledge. While independently astute, constructive and often practical, these research studies have not examined the extent to which online collaboration can contribute to non-formal learning initiatives such as the volunteering practice of knowledge workers. Book highlights the patterns of use of relevant web-based collaboration tools among the volunteer employees to create and sustain a sense of online community. It is clear that when it comes to online communities there is a mixture of entanglements, partnerships, negotiations, and resistances between these tools and human actors. As Waltz (2006, p. 56) states "these tools require new ways of usage as new methods for interaction among uses are being discovered". These technologies are significant, given how they create new methods for understanding, working, and training. As many people, including employees, are venturing into online communities of all shapes and sizes attending to these experiences becomes crucial.

Despite the vast amount of existing literature on employee volunteering programs (Hess et al., 2002; Warner, 2004); it should also be taken into account that apart from the conventional types of workplace learning, there is an excess of **informal learning** opportunities that equip individuals with the related professional and personal competencies required to manage the difficulties they encounter throughout their personal and work lives. Nevertheless, there is a shortage of studies with a focus on the voice of the individual employee while the tool or platform design seems to have gained more attention. So, especially given the popularity of Web 2.0 tools, there is a need to investigate employee's current experiences and expectations of these tools across the broad range of adult, community and work-based learning (Creanor et al. 2006).

The study essentially examines both the strategic use of online collaboration tools and its perceived benefits among volunteers who are situated in the practice (Salomon, 1993) of volunteering while working together with the aim of sharing and enhancing their expertise. This also contributes to the understanding of the use of collaboration tools and shapes the volunteering practice that takes place. The emergence of the collaboration patterns reveal the extension of workplace learning on a non-formal level within the context of volunteering which embeds both knowledge discovery and construction. I also reflect upon various approaches for leveraging the existing digital tools for facilitating knowledge creation and appropriation of content material for the purpose of collaboration.

The study aims to reveal the usage patterns of online collaboration tools by employee volunteers to contribute to their online and collaborative interactions. My focus is on the affordances and perceived benefits of the tools that potentially make the volunteer's use of, and decision to use, the technology differently. By seeking theoretical insights on how to utilize these tools throughout the volunteering process, my aim is to serve those people who are involved in not only similar **professional development** or non-formal learning programs, but also interested in using online collaboration tools in general and do research on them.

The volunteering program includes an informal curriculum to be completed by teams made up of 8–12 employees who were sent to developing countries to participate in local social and economic development projects, based on their individual expertise. The reasons for selecting such an employee volunteering program as a case study are linked to different benefits:

1. Book reviews a range of online collaboration tools and studies how they have been utilized in supporting employee volunteers for sharing their knowledge and expertise during their volunteering practice. These specific tools have rarely been used for an employee volunteering until recently, and there may, therefore, be value in exploring the potential role that online collaboration tools might play in the practice of employee volunteering and eventually the process of non-formal learning. The in-depth case study raises an awareness of the contribution of online collaboration tools for the practice of employee volunteering; highlights potential barriers to and facilitators of online collaboration and offers methods and approaches for using related tools for other ways of non-formal learning. Such an approach could be significant in terms of seeking to make online collaboration as much a critical issue for the new trend of employee development programs.
2. Book also has a secondary objective which is to contribute to the existing list of methodologies based on participatory design for researching about the digital experiences of employees with new technologies. Employee volunteers in this study have not only been involved as research subjects, but also as partners whereas they contributed to the identification and reframing of the research questions. Based on the data analysis the results have been put under scrutiny in collaboration with them.

## 1.2 Asking the How Question

One of the organizations developing learning programs on a global scale is IBM. Since 2008, IBM has been implementing a volunteering program referred to as **CSCs** which includes an informal curriculum to be completed by teams made up of 8–12 employees who were sent to developing countries to participate in local social and economic development projects, based on their individual expertise. (see Appendix E). IBM volunteers are recruited from different countries and these professionals prepare 3 months in advance, focus on their project for 1 month, and offer then consulting services and convey their knowledge in a structured approach. During the program, it is ensured that the volunteering project really integrates with local communities and creates an impact. Building strong relationships with local communities or government officials is also important as this will ultimately benefit the business as well.

Participating cities not only receive free consulting services, but also make use of technology to solve their issues. Apart from building strong relationships

6  *Introduction*

with communities, IBM differentiates itself from other companies in the technology sector by contributing to the less developed societies through this volunteering program. The CSC program has also won various rewards in the field of corporate volunteerism.

IBM considers the integration of online collaboration tools into this volunteering program to be a seminal process. It views the use of these tools throughout the program as a way to fundamentally shift how employees work together and can transform the volunteering process.

One of the interesting features of this CSC Program is its emphasis on the use of online collaboration tools throughout the volunteering process. Given the technology literacy skills of IBM employees, this might not be surprising, yet there are important lessons to learn for companies being involved in other sectors as well, as there is an increasing emphasis on the skill development of employees through various volunteering programs (Golensky, 2000). Additionally, relevant success factors and barriers with regard to the use of these online tools can be noticed and some suggestions related to the new trend of employee volunteering can be collected.

Before I engaged in the study I was working as a program manager for one of the implementation partners of this CSC Program (UNDP Turkey) in my home country Turkey. My own status as an insider, a former employee in one of the stakeholders of the CSC Program involved in the study, will serve as a strong asset. As an insider, I understand the culture of the organizations involved in the study and easily gained access to the interviewees. However, I am also aware that the researcher's background and related professional experience also have a potential effect on the interpretation of data (Patton, 1980, 1990).

## 1.3 Asking the What Question

The main purpose of book is to provide insights into how online engagement enabled the continuation of traditional workplace learning practices such as volunteering and opened up possibilities for new ways to contribute to the learning process of employees. However, these digital spaces are not devoid of any issues and hence suggest the need for employees to be conscious of the emerging issues. The underlying objectives of the research study are:

1. Explore the collaborative online interactions of employee volunteers by investigating their ways of knowledge production by utilizing digital tools and identity factors that provide or limit their joint activities with these digital tools.

2. Make suggestions and raise important issues for organizations interested in non-formal learning practices within such fluid settings of their employees grounded in my conceptualization of the choices and expectations of the employees.
3. Make a contribution to the existing list of methodologies based on participatory design for researching about the digital experiences of employees with new technologies.

For this purpose, the study aims to provide answers for the following research questions:

1. How are online collaboration tools used for the volunteering practice of knowledge workers?
2. What are the assumptions of knowledge workers about the benefits and challenges in using these tools for such a practice?

Collaboration is getting involved in a joint activity to create a shared new or greater understanding about a process, product or an event that one would not be able to come to on his own (Schrage, 1990). Online collaboration tools refer to the web-based tools such as **social networking** sites, wikis, blogs, or discussion forums utilized by various individuals to accomplish a common task.

Weick (1990) describes technologies as equivocal, using the phrase "technology as equivoque". He defines an equivoque as being subject to different possible interpretations and therefore appearing to be complex and uncertain (Weick, 1990). As technologies are equivocal, different meanings can be attached to them. So, there is certain value in exploring the potential role that online collaboration tools might play in the development of volunteering practices.

# 2

# Some Theories about Online Collaboration

As the theories provide us with lenses to approach a real-life phenomenon, they definitely should be mentioned although it may sound boring to the reader at first glance. The purpose of this literature review is to position this project within the large field of 'workplace learning', first by discussing some issues that are of relevance to this case, and thereafter, by going more thoroughly into the particular trend that is the concern of this study, 'online collaboration tools'. The two streams of research reviewed for this study are workplace learning and **CSCL**.

For me, research is a form of communication. Whether psychological, sociological, or in natural science, research can be defined as a kind of discourse producing narratives with which other human practices can be modified and gradually improved. Such a view is also consonant with the assumption that discourse plays a constitutive role in all human practices as stated by Vygotsky (1978). In order to place this discourse about online collaboration into the broader context of the volunteering practice that can be regarded as an informal type of learning, the chapter begins by introducing two case relevant aspects of workplace learning (Section 2.1). Next, Section 2.2 gives attention to the social aspects of workplace learning. Of course, anyone else might prefer to include other fields of study in a discussion about employee volunteering; yet as this study focuses on the use of the digital collaboration tools these two fields are mostly important to look at given the limited scope.

The term "workplace learning" does not refer to a single concept. Rooted in adult education, the term has frequent links to formal education, and a clear focus on the individual learner (Elkjaer and Wahlgren, 2006). Although this type of learning is often related to the narrow vocational training paradigm 'to make the employees fit for the job', the approach also has a broad societal perspective such as employee volunteering programs. The learning is viewed from the employees' angle, with an emphasis on general personal development

(Illeris et al., 2004) which is also one of the underlying goals of the CSC Program. In line with the growing interest in workplace learning, informal learning has increasingly been acknowledged as a crucial ingredient of the concept (Elkjaer and Wahlgren, 2006). In this book, workplace learning will be used as a generic term for learning and competence development through various means such as employee volunteering programs like CSC.

The last section of the literature review provides an overview of online collaboration tools with a focus on the more popular collaborative technologies as these tools have been widely used throughout the CSC Program. Before delving into the literature on CSCL, an in-depth explanation of the terms "collaboration" and "**cooperation**" are also provided as there is still an ongoing debate between these two terms.

## 2.1 Workplace Learning

Any volunteering practice of employees has the underlying aim to encourage informal or non-formal learning and improve their professional skills. As such, it is an important aspect of workplace learning. Workplace learning is nothing new. Even during the early stages of the formal education system, skills regarding specific work tasks were conveyed from one generation to the next (Rosenberg, 2006).

The industrial revolution and assembly-line production represented a break with this old rule. From then on, the work was organized according to the requirements of the *production technology* (Illeris et al., 2004). Because of a growing demand for specialized knowledge, more differentiated qualifications, and an increasing social complexity, skill acquisition, and programs for education later moved from the workplace to off-site course activities and schools. This tendency towards learning, as something *localized outside the workplace and separated from the work situation* continued during large parts of the twentieth century. At the expense of learning at work, traditional apprehension toward job-related learning at school increased, and learning was disconnected from the work situation (Illeris et al., 2004). However, from the late 1980s and at the beginning of the 1990s, a new school of learning and competence development appeared. Due to emerging organizational needs for handling increased competition, a growing specialization, a continuous development of new technology and a number of rapid and little predictable changes, learning again returned to the workplace (Rosenberg, 2006). In large companies this trend started as early as in the 1960s, with in-house mass education to increase the workplace performance (Rosenberg, 2006).

Based on the idea that the effectiveness of learning depends on the degree of the matching between the learning context and the use context (Clarke, 2004), this new approach proposed to integrate the learning process with the job-specific competence needs. One of the two most frequent models of this type are the course-based with job-relevant preparation and supplementary work tasks, the other one is organized as continuous on-the-job-training.

In order to understand the CSC Program clearly, I had to familiarize myself with a number of taxonomies existing in the WPL as well as in the management, human capital and organizational literature (OL; Boud and Solomon, 2001; Malcolm et al., 2003). However, the taxonomies from the two traditions differ both in scope and focus, primarily in alignment with the underlying tradition on which they are grounded. Nevertheless, both traditions mainly focus on the *individual* as the learner (Elkjaer and Wahlgren, 2006, p. 2).

To provide a short overview of these two traditions, with its roots in adult learning, WPL often makes a distinction between formal, non-formal and informal learning (Billet, 2001). Other categories, such as guided learning (Billet, 2001) and incidental learning (Marsick & Watkins, 1990), are also used. Because the categories often slide into each other and can be combined in different ways, the distinction between the different categories is to some extent unclear. This is also the case with the three categories of formal, informal and non-formal learning. Based on an in-depth literature review and empirical studies in this field, Malcolm et al. (2003) concluded that when it comes to making a definition of formal, non-formal and informal learning the existing literature has neither unified definitions commonly agreed upon for each of these terms, nor draws exact lines between them. Instead of using these concepts, they therefore suggest characterizing workplace learning according to four dimensions: the process; location & setting; purpose; and content.

In contrast to WPL, OL draws on theories of **organizational learning** and theories of management (Elkjaer and Wahlgren, 2006). Based on these theories, as well as on practice-based approaches, learning at work, during the past two decades, has been conceptualized through paradigms like "**organizational learning**" (Argyris and Schön, 1996), "knowledge management" (Nonaka and Takeuchi, 1995) and involvement in "**communities of practice**" (Wenger, 2000, 2004). What the three conceptions have in common is a view on workplace learning primarily as a collective phenomenon (Miettinen and Virkkunen, 2006), or in Elkjaer and Wahlgren's (2006, p. 29) words "as a development from the individual as a container and processor of information and knowledge by way of the individual as oriented towards personal development towards development of membership and professional identity".

Only a few taxonomies make an attempt to integrate the two traditions. One such consists of the following four categories: *formal; informal; individual;* and *organizational* learning (Rosenberg, 2006). Another taxonomy is offered by Illeris et al. (2004, p. 139), who distinguish between: (i) the direct approaches having the objective of the measurement of learning, and (ii) the initiatives that enrich opportunities for unplanned learning through efforts of turning the workplace into a learning environment. While the first one is manifested at its most extreme by the educational activities in the WPL tradition, the second one is manifested through paradigms such as "the learning organization" (Argyris and Schön, 1996).

The incompatibility of the taxonomies makes it rather difficult to give an exact a priori categorization of IBM's CSC Program. Using aforementioned categories, I prefer to describe the CSC Program as both non-*formal* and *collective*. Non-formal, in that the learning process was not "stimulated by planned and systematically implemented training efforts, for instance courses and participation mostly occurs on a voluntary basis rather than in a prescribed manner; and collective, in that the individual employee, as a result of the learning activity, was not only expected to be able to develop or change his competence, but eventually also to modify current work patterns and/or develop new patterns (Rosenberg, 2006).

## 2.2 The Social Aspects of Workplace Learning

It has been more than ten years since there has been a shift from the individual to the constructive and social aspect of knowledge in the existing epistemologies (Easterby-Smith and Lyles, 2003). Such a direct shift of focus onto the social nature of meaning and practice can result in the redefinition of the organization itself as a community of practice (CoP), with organizational dimensions that convey meaning to these practices.

The prominent scholars Lave and Wenger (1991) who firstly made a definition of CoP in their famous book with the title *"Situated Learning: Legitimate Peripheral Participation"* studied how situated learning takes place as a result of the relationships built by "master practitioners" and "newcomers". CoP's can also refer to places in which "communicative action" occurs (Polanyi, 2002). The mutual creation of knowledge mediates these actions (Wenger, 2004). While CoP's function as a ground for knowledge creation and transfer (Lesser and Prusak, 2000; Wenger, 2004; Wenger and Snyder, 2000) they exist at the crossroads of intellectual and social capital. Within the current body of literature it is a common belief among scholars

## 2.2 The Social Aspects of Workplace Learning 13

that CoP's support the basis of social capital, which is mandatory for creating knowledge and its dissemination (Lesser and Prusak, 2000, p. 124).

According to Wenger (1999, p. 6, 7), CoP framework can be implemented within both "intra" and "inter" dimensions of organizational settings due to being "an integral part of our daily lives". Building further upon the concept of CoP, Wenger (1999; 2000; 2004) utilized it to establish a comprehensive theory of how individuals within collective settings such as organizations work together. In his book, *Communities of Practice: Learning, Meaning, and Identity*, Wenger (1998, 1999, p. 30) states that organizations can be considered as assemblies of CoP which can reach even beyond their confines and be situated either within or between formal networks. In addition, some scholars utilized the concept of CoP to put cross-sector collaborations under scrutiny (Lathlean and le May, 2002; Dewhurst and Navarro, 2004). These scholars have also contributed to my motivation for approaching the CSC Program from the perspective of CoP. These studies suggest that organizational initiatives provide a fruitful ground to implement the CoP theory.

The term "joint enterprise," referred to as the shared purpose of practitioners in a particular field is used as one of the main characteristics of a CoP (Wenger and Synder, 2000). Similarly, according to O'Donnell et al (2003), CoP's are formed around a common interest established upon the values of their members. These shared interests are set into a negotiation on a communal basis (Wenger, 1998, p. 78) around a common purpose. Wenger (1998, p. 51) describes a "practice as a process by which meaning is provided for one's engagement within the world". According to Wenger (1998, p. 81), "mutual accountability", which refers to the degree of reciprocal relationship among its members, acts as a glue in terms of holding these joint enterprises together. The "shared repertoire" is another feature underpinning CoP (Wenger, 1999, p. 82) and this "shared repertoire" includes the tools and techniques in order for negotiating the meaning and making learning happen (Wenger, 1999). Possible forms for this repertoire range from an informal discussion during a coffee break to a structured meeting based on some decision-making criteria. According to Wenger and Synder (2000), as CoP's often have connotations to business units or teams; additional effort is required to integrate them into organizational settings in order for their power to be realized.

IBM's CSC employees can be considered as communities of voluntary practitioners and their means of communication should also be taken into account. These means of communication range from face-to-face interactions to the use of various digital tools. In other words, it is not sufficient to focus only on the individual elements of the CSC Program such as the volunteers or online collaboration tools, but in particular on their mutual interplay. Crossan et al. (1999) states that one of the main barriers against theory development with regard to any organizational practice is whether the unit of analysis should be individual, group, organizational, and/or inter-organizational. Furthermore, some theorists assert that an organizational practice would not be complete without the sharing of information and the development of common meaning (Daft and Weick, 1966; Huber, 1991; Stata, 1989). Consequently, as an organizational practice must be shared and integrated with the learning done by others (Brown, 1993; Daft and Huber, 1987; Daft and Weick, 1966) the unit of analysis should be the group. Other scholars assert that the unit of analysis should be the organization itself as much needs to be done by organizations themselves due to the fact that the activity is stored with organizational structures, procedures or systems (Duncan and Weiss, 1979; Hedberg, 1981; Shrivastava, 1983; Fiol and Lyles, 1985; Levitt and March, 1988; Stata, 1989; Huber, 1991; Chi-Sum et al., 2008). By taking into account these theoretical perspectives, the unit of analysis of this study will be the group as it focuses on the different CSC groups made up of IBM employee volunteers.

### 2.2.1 Summary for WPL

One of the major strengths of the existing body of workplace literature is the consideration of organizations as assemblies of CoP that can go beyond their physical boundaries and be situated either within formal networks (Wenger, 1998). Taking a similar approach, this study considers CSC employees as communities of voluntary practitioners embedded within a learning network.

Another strength of workplace literature is that there is a common agreement that studies mostly fall into one of these two camps as indicated by Illeris et al: (2004): (i) studies with the direct approaches having the objective of the measurement of learning, and (ii) research initiatives that enrich opportunities for unplanned learning through efforts of turning the workplace into a learning environment. As the CSC Program offers a unique environment for unplanned learning, book further contributes to the literature of workplace learning.

One of the main weaknesses of the literature is that there are several taxonomies used which differ from each other both in scope and focus. Although most of these studies focus on the individual as a learner, this study goes one step further by providing insights about the mutual interplay of IBM employees and the digital collaboration tools used throughout an employee volunteering program. These tools, as their means of communication, range from face-to-face interactions to the use of various digital tools. In addition, no exact lines are drawn between the terms of formal, non-formal, and informal learning. By framing the CSC Program within the context of non-formal learning practice, this study contributes to a clearer distinction between these terms.

## 2.3 Exploring CSCL in Depth

Any academic discussion of online collaboration involves the practice and theory of CSCL. While the focus of much current CSCL work with regard to workplace learning is rooted in workplace interaction, we should keep in mind that contrary to popular belief, CSCL could especially make a difference when it comes to learning outside the boundaries of organizational settings. So, apart from the daily work practices of individuals, the social "situatedness" of learning (Winograd and Flores, 1986) should also become the focus of these discussions (Lave, 1988). Due to the adoption of such an alternative approach "outside-class" activities are considered as a crucial aspect of the social background with regard to the process of learning (Cole and Griffin, 1987).

From the theoretical perspectives of CSCL, learning should be assessed on the group level while technology can support the group processes: According to Scardamalia and Bereiter (1996), the community learns as a whole in a computer-supported learning community while the term "community" itself needs a re-conceptualization taking into account the definition provided by Lave and Wenger (1991). Engeström (1999) took a wider learning approach and studied how learning occurs during the interaction of multiple groups among each other. Stahl (2001) claims that these theoreticians (e.g., Lave, 1996; Engeström, 1999) derive their social theories based on Hegel (1967), Marx (1976), and Vygotsky (1978) and that these CSCL theories are disputative due to the increasing complexity of the history of philosophy and theory since the times of Descartes. According to Kant (1787), our conceptualization of the outer world was represented by the human mind, which involves a basic structure rather than being simply given by the material world. Hegel (1807) adopted a developmental view and grounded the process of representation in changes throughout the history. According to Marx (1867/1976), these changes are grounded within socio-economic phenomena. Later on, another famous scholar, namely Heidegger (1927), suggested another perspective in which the human-being is more firmly situated in the world than Descartes' approach. Figure 2.1 shows a graphical representation with regard to the different social and individual theories of learning.

## 16   Some Theories about Online Collaboration

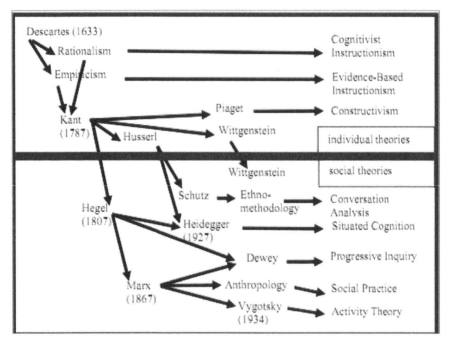

**Figure 2.1**   Graphical representation of individual and social theories of learning (Stahl, 2001).

Taking into account these individual and social theories of learning there are two main approaches of defining CSCL:

1. According to the first perspective, CSCL can be seen as an "umbrella term" which provides a fertile ground with regard to the development of multi-faceted perspectives on related topics. In fact, this approach provides a further ground for the creation of many new research areas such as **CSCW** (Bannon et al., 1988; Bannon and Schmidt, 1991).
2. The second perspective is related to understanding the related problems and concerns in detail and establishing a shared understanding on the object of study which would further contribute to the development of the field. As there is no unified definition for CSCL, a compositional perspective might be taken in which the meaning of the term is built from its components. So, possible questions that can be asked include what do people mean by collaboration or learning and by CSCL. Rather than imposing an exclusive interpretation on the meaning of CSCL, the focus of research can shift to workplace learning, in this case to the specific

initiative of employee volunteering, and how it might be supported by the online collaboration tools.

I will make an effort is made to provide operational definitions related to various online collaboration tools. First, I take a close scrutiny at the keywords such as collaboration or cooperation as different writers are making different uses of the same words, and such different uses are likely to be grounded in differing premises about the nature and source of the phenomena.

### 2.3.1 A Crucial Definition: Cooperation versus Collaboration

The Latin root for the word "collaboration" implies the process of coming together for work, while the word "cooperation" focuses on the output of this work (Myers, 1991). Based on the philosophical writings of Dewey (1938) co-operative learning is a more popular term in US and it stresses the social nature of learning and the work on group dynamics.

Both types of learning are grounded in constructivist theory (Schrage, 1990). According to this theory, knowledge is reconstructed by learners and expands through new learning experiences which come about through transactions and dialogue among individuals in a social setting (Schrage, 1990). So, individuals gain an understanding of different perspectives through a dialogue with others.

In a detailed discussion of the different meanings of cooperative and collaborative learning, Dillenbourg (1999) stated that while in cooperation the work is divided among the related group members so that sub-tasks can be completed individually and eventually the final output is produced, collaboration implies coming together to accomplish a common task.

Another difference in the meaning of these two terms has been pointed out by Roschelle and Teasley (1995). These authors asserted that collaboration implies a process of negotiation and sharing of meaning involved with regard to the task to be completed. According to these scholars, collaboration is "a synchronous activity as a result of an ongoing attempt to create a shared conception of a problem". (p. 70). So, in addition to the participation of individuals as group members, the creation and sustainment of shared interpretations of tasks undertaken in a mutual manner in group settings are also crucial for collaborative learning.

Schrage (1990) defined collaboration as an act of shared creation and/or shared discovery. According to him, increased communication cannot substitute for increased collaboration. Increased information does not necessarily improve thinking; creating a shared understanding is different from exchanging information. According to Schrage (1990), collaboration refers to interaction with the purpose of creating a shared new or greater concept about a process, product or an event that one could not come up with on his own (Schrage, 1990). Seen from this perspective, cooperation is a crucial factor for collaboration and the use of term 'collaboration' might sound more relevant during the CSC Program. Rather than going off to do things individually, participants remained engaged with a shared task that is produced and maintained by and for the group (Roschelle and Teasley, 1995).

On the other hand, in order for cooperation to occur, individuals make their personal contributions and refer to the set of these individual outputs as their collective product. Within the context of cooperative groups, learning is recognized as an individual event in contrast to the collaborative learning which views learning as a social incidence leading to mutual construction of knowledge (Roschelle and Teasley, 1995).

Another scholar who was interested in the difference between these two terms was Rockwood (1995). Rockwood (1995) agrees with the similarities of these two terms as both terms relate to the use of groups and assignment of specific tasks. Rockwood (1995) claims that while cooperative learning is related to canonical (traditional) knowledge collaborative learning is based on the social constructivist movement. These statements are in alignment with Schrage's (1990) assertion with regard to the shift from "cognitive (foundational) understanding of knowledge", to a non-foundational ground where knowledge is understood to be a social construct and learning to be a social process".

In order to avoid to create a 'we versus them' mentality while pointing to these differences between the two terms we should acknowledge various benefits underpinning both ideas rather than disregarding any advantage caused by the interactions created by both methods. Yet, as the underlying philosophy of collaboration is willingness for accepting the responsibility and sharing the authority I prefer to use the term "collaboration" throughout book. What is important here is to have an awareness of the vast differences in meanings among scholars when using these terms rather than to come up with a final definition of these terms.

### 2.3.1.1 Collaborative learning
Due to the different theoretical perspectives adopted and research paradigms used in studies about collaborative learning, taking a holistic approach on these studies is often times difficult. Some researchers prefer to derive their theoretical framework from the studies of such scholars like the Russian

scholar Vygotsky (1978) specialized in socio-cultural psychology. One of the important concepts in this field of study is the notion of the "zone of proximal development" (ZPD) as referred to by Vygotsky for the space in which learning takes place. This realm refers to the gap between the existing stage of development which is specified by means of individual problem-solving and the potential stage of development which is specified through solving problems with the support of an adult or more capable peers (Vygotsky, 1978). This concept deserves much attention not because of the fact that it has been subject to many educational research in which the computer has been used as a medium (Newman et al. 1989), yet as it views learning as a collaborative process. Most of the studies do not address specifically whether a single person is collaborating with someone else or with a computer system. In some other studies, the subjects might be classmates, who had common experiences together or as teams who put effort for improving their actions via means of the computer over a span of time. So, all these different kinds of interactions might lead to different degrees of collaboration. Within the up-to-dated body of literature, a framework that has been commonly agreed upon in terms of making a comparison and contrast among these studies about collaborative learning does not exist yet. In my point of view, such a framework should include the following aspects as an example: The nature of the collaborative task, the number of collaborators, the relationship among collaborators, the underlying motivation for collaboration, the context of collaboration, the means of collaboration: whether it is computer-mediated or not, as well as the duration of collaboration.

Despite the different approaches used in CSCL studies one commonly agreed aspect of learning is its situatedness in 'peer networks' (Brown and Duguid, 2000; Hiltz, 1990). Brown and Duguid (2000) claim that the balance between face-to-face interaction and computer interaction is a critical factor for the successful existence of an online community. These authors also assert that when it comes to more complex and geographically distributed virtual environments this balance becomes even more important (Brown and Duguid, 2000).

There are various theoretical perspectives with regard to the use of online collaboration tools. To begin with, Kurland and Kurland (1987) make a reference to Collins (1984) in one of their research studies in terms of the ways the computer can be used to contribute to the process of learning. The computer does not only provide an opportunity to replicate situations that would be difficult to be simulated in the real world, but it also offers the possibility to keep track of learner actions which can be useful for improving the problem-solving strategies of learners. Moreover, by making the process of thinking visible, it can facilitate functional learning environments which support the learner in pursuing meaningful goals and acquiring these goals (Kurland and Kurland, 1987). Such a passive role of computers is similar to the use of other possible tools such as pencil and paper which in my point of view does not very much relate to the use of online collaboration tools implied by the process of employee volunteering.

Still another viewpoint views the computer as a kind of "tutor" with whom the learner either interacts or collaborates. Two kinds of research activities follow this direction. The first one with a focus on more traditional CAI or CAL is related to "drill-and-practice" type exercises which are regarded as not very intelligent and hence no longer seen as having a crucial role in the context of "collaborative learning". The second one involves the Intelligent CAI, which regards technology as a coach that is able to support the user in his conceptualization of a particular area by showing the gaps or mistakes in the learners' conceptual model evidenced based on their responses to problems (Sleeman and Brown, 1982). While the latter one has a stronger connotation to "collaboration", there is the implicit assumption that the computer could somehow take the place of the teacher.

Leaving these two main viewpoints aside, I would support a third point of view. According to this view, technology is used as a medium for collaborative learning as this supports individuals for their communication or sharing tasks on joint activities. Due to this intermediary role of the technology the focus shifts to its use as a medium through which technology provides support for individuals being in collaboration with each other. Again, rather than trying to come up with a correct perspective, effort should be given onto understanding the particular perspective of taken by the research study conducted. By keeping this in mind a more coherent assessment of "success" or "failure" of the research study can be made.

Also, it is important to keep in mind the three main aspects of collaboration as Cogburn (2002) suggested; namely: collabor- ation readiness, collaboration infrastructure readiness, and collaboration technology readiness (Cogburn, 2002).

- Collaboration readiness: Readiness to collaborate or experience with collaboration is certainly a requirement for a successful collaboration. In CSC Program, IBM volunteers had professional experience with the use of digital collaboration tools. Their field work had pre-determined rules with regard to the sharing of information and creation of online knowledge resources. This community of volunteers was "collaboration

ready". Often the major component of this aspect is a real need for working together in order to achieve a common goal (Cogburn, 2002). This aspect has the following components: motivation for collaboration, shared principles of collaboration, and experience with the particular units of collaboration (Cogburn, 2002).
- Collaboration infrastructure readiness: There is a need for adequate infrastructure for the proper operation of the modern collaboration tools. Needless to say, bandwidth intensive tools such as video or audio need adequate networking. With regard to the technology infrastructure such as currently updated network environments there should be no issue for off-the-shelf applications to run effectively (Cogburn, 2002).
- Collaboration technology readiness: Although collaboration technologies have been available for almost two decades, they have gained more attention in the mid-90s. To give an example, even though e-mail which can be considered as a simple form of collaboration technology was first used in the 1960s, its use become widespread during the 1990s (Cogburn, 2002). Due to the fact that there may be a need for an advanced level of training for a more complicated technology it is suggested to assess the state of technology readiness in the related team or group to ensure its success (Cogburn, 2002).

### 2.3.2 Collaborative Learning Tools

As the CSC volunteering program requires the volunteer employees to work with others in teams while learning collaboratively the related literatures on both distributed collaborative and distance-independent learning are also to be explored for book. Johansen (1988) defines the **collaborative learning environment** as interactive pedagogical approaches that utilize web-based technologies for the creation of an infrastructure to fulfill both synchronous and asynchronous requirements of distributed learning teams.

Table 2.1 provides the differences between traditional and collaborative learning as explained by Mandl and Krause (2001). A constructivist learning theory underpins the concept of CSCL. According to this theoretical approach, learning involves a process guided on one's own which requires a conscious knowledge creation and hence the previous experiences, skill set and mindset of the individual influence this process (Mandl and Krause, 2001). Additionally, there is a second constructivist approach with regard to knowledge-sharing: "to solve problems in a self-organized way" (Arnold and Schussler, 1998, p. 78). Within this regard it is crucial for organizational

**Table 2.1** Differences between the traditional and collaborative **e-learning** model (Mandl and Krause, 2001)

|  | Traditional Approach | Collaborative Approach |
| --- | --- | --- |
| The objective of learning | Being qualified for expertise | Skill |
| Know-how | In progress, memorized | Construed |
| Paradigm | Solving a problem, gaining an understanding | To enhance-related experiences and practices |
| Technology use | Dissemination | Communication, learning in collaboration |
| The mode of involvement for learner | AM | PM |
| Interaction type | Delivery model | Dynamic and complex model |

stakeholders that different types of learning are facilitated by supporting learner-oriented, social and situative learning (Mandl and Krause, 2001).

Timothy Koschmann (1996), one of the prominent scholars in this field asserted that this shift in pedagogical models due to the use of technology represents the start of a new paradigm according to the Kuhnian perspective. Koschmann (1996) further stated that within the context of CSCL, the emphasis shifts from the personal development onto the group cognition and due to the incompatibility of this perspective with the more individualistic conventional view, it is in need of a new paradigm as determined by Kuhn (1962).

In a similar vein, Sfard (1998) referred to the incommensurability of these two paradigms using both AM and the PM. In empirical research using the AM, the focus is on the change of individual mental models as learning involves conveying knowledge to the user. On the other hand, PM views the learning process as being more social and hence taking place among different subjects on a group level. So, the emphasis within the experimental research, therefore, is on the motives of participation within groups. Rather than identifying a change in paradigm, Sfard (1998) gives equal weight to both metaphors.

Stahl (2001) stated that due to the strongly held Western views based on Descartes' "cogito ergo sum" a shift from an individualistic model into a group-based cognition did not take place yet. Given the widespread use of social and digital collaboration technologies in our times Stahl's perspective may no longer be relevant. However, Stahl (2001) also suggests the reinforcement of a strong group participation which is relevant for book due to the different participation levels in the CSC Program. Another prominent scholar in this field is Maxwell (2002) who argued that the condition of incompatibility has not (yet) been fulfilled, so Maxwell (2002) is rather uncertain about the change of the existing paradigm and stated that a shift from one paradigm into another one has not occurred yet. Maxwell (2002) preferred to confirm that a change rather than a paradigm shift is in place and he put this change into analysis from a pragmatic perspective and stated that they all belong to the same genre who all have the same justification to exist and develop (Maxwell, 2002).

In my view, rather than trying to come up with a unified approach for empirical research in CSCL researchers should focus on how individuals collaborate with digital tools which might also be relevant for the CSC Program. In my point of view, the aim should be to elaborate on the ways of using these tools in an effective way in order to obtain the commonly shared goals which is one of the underlying features of collaboration.

In the next sections, I provide some examples of applications and tools as well as their uses in an attempt to contribute to collaborative learning within the context of employee volunteering.

### 2.3.3 Groupware

**Lotus Notes**, which is classified as a **groupware**, is used within IBM to a great extent. Ellis et al. (1992) make the definition of groupware as applications which are based on the computer technology and which provide the team members working on a common task (or goal) through the means of an interface to a shared environment (Ellis et al., 1992). Groupware can be described as an organizing framework for sharing information within a workgroup by facilitating structured communication among group members and coordinating group activities (Bock and Marca, 1995; Flate et al., 2011). The software is designed and implemented to specifically support interaction (Zack and McKenney, 1995).

Schrage (1990) believes that the basic premise underlying groupware is an increased requirement for shared effort, cooperation and collaboration. Schrage (1990) states that the goal of groupware is to create value through human interaction rather than only providing information.

Groupware also includes company-or department-specific applications, e.g., contact management, and workflow process management. Collaborative technologies augment decision-making process by permitting and enabling idea generation or other courses of action (Neilson, 1997; Caballé, 2011), so this distinction between groupware and collaborative technologies is important. In this study, groupware is used to describe traditional technologies that have been commonly embraced by most companies such as e-mail. The focus of the study is on collaborative technologies, the less ubiquitous tools that support cooperative work such as popular Web 2.0 tools like blogs, wikis, discussion forums and synchronous chat.

One of the main groupware technologies used at IBM is Lotus Notes which was developed and introduced in 1989 as a groupware tool to enable functions of communication, coordination, and collaboration. There are a few empirical studies about the actual uses of groupware (Ciborra, 1996; Perez et al., 2011). In an early study, Orlikowski (1992) explores the introduction of Lotus Notes at a large professional services organization in order to obtain a conceptualization of the changes affecting working patterns and social interaction realized via use of technology. According to Orlikowski (1992), if the value of the collaborative nature of groupware is underestimated then these technologies will not be utilized to their full potential. Moreover, when the premises underpinning this technology such as collaboration or shared effort run counter to an organization's hierarchy, structure or culture then the groupware technology will not provide any value for its collective use (Orlikowksi, 1992). Orlikowski (1992) concludes that groupware implementation does not easily fulfill the potential for enhanced organizational effectiveness and that usage and satisfaction might change over time based on experience (Orlikowski, 1992; Romero, 2011). Darr and Goodman (1995) also assert that time is a major variable in self-reported use and satisfaction with Lotus Notes.

Few researchers focused on whether work group computing is enhanced communication or whether it provides for true collaboration. Various studies have been undertaken with regard to the use of electronic mail (Finholt and Sproull, 1990; Zack and McKenney, 1995; Hopkins, 2011), yet there are not so many published studies about groupware technologies that involve more collaborative features with the exceptions of Darr and Goodman (1995) and Orlikowski (1992).

Hiltz and Johnson (1989) and Lou (1995) present quantitative studies that researched users' acceptance of groupware, identifying use, satisfaction, and benefits as dimensions for acceptance. Lou's study of the Notes technology presented statistically significant results, yet the coefficients of determination were at low levels; Lou (1995) attributed this to factors such as user experience and organizational context. Similarly, Darr and Goodman (1995) evaluated the usage functions for communication, coordination and collaboration. The hypobook was that individuals who participate on project teams would report higher Notes usage for collaboration than people who did not participate would on project teams. Actual findings were that communication usage is greater than collaboration and that collaboration is the least used function (Darr and Goodman, 1995). Contrary to this hypobook, organizational culture and time since adoption rather than participation on a project proved to be strong factors in the use of Notes for collaboration. Darr and Goodman (1995) also found that the extent to which an organization is team based did not influence the frequency of Notes use.

### 2.3.4 New Generation of Collaborative Learning Tools

One of the reasons that there has been an increased interest in using technology to enhance individual and group work processes is that personal computing and networks have become ubiquitous (Sproull and Kiesler, 1995). The second generation of web-based applications, recently described as "Web 2.0" (O'Reilly, 2005, 2007) have become so popular that a third generation of these tools is already on the rise. I will refer to these Web 2.0 applications as the new generation of collaborative learning tools. Indeed, traditional **learning management systems** (LMS) have now embedded a much larger number of various applications based on Web 2.0 ranging from wikis and feeds of **RSS** to **podcasts**, social media, and peer-to-peer file sharing systems, and many more.

Despite its wide adoption of Web 2.0 applications, confusion about the term "Web 2.0" still exists among its users. Despite the widely held belief that 'Web 2.0'is a meaningless marketing phrase it is seen as providing a new conventional insight (O'Reilley, 2005; Guitert, 2011). Scholars in this field take different stances with regard to this new trend.

To begin with, according to O'Reilly (2005), while Web 2.0 refers to the network as platform dispersed over all connected devices, Web 2.0 applications refer to those applications which have embedded the inherent benefits of that platform: application gets updated continually the longer it is being used and data is amalgamated from multiple sources by individual users who eventually create their own web services with the purpose of building a network through an 'architecture of participation' resulting in rich user experiences (Figure 2.2; O'Reilly, 2005).

Similarly, Anderson (2007) refers to Web 2.0 as a group of popular technologies which have become widespread such as blogs and wikis that increases the social connections on the Web so that anyone is able to contribute or edit information on the Web. Most of these applications have already become very mature due to their common adoption by a large number of

| WEB 1.0 | ➡ | WEB 2.0 |
|---|---|---|
| Encyclopaedia<br>Personal Web Sites<br>Publishing | | Wikipedia<br>Blogs<br>Participation<br>Syndication (RSS) |

**Figure 2.2** The differences between **Web 1.0** and Web 2.0 (O'Reilley, 2005).

users and due this increasing number of users, these tools are being improved with new features and capabilities on an ongoing basis (Anderson, 2007).

What this definition implies is that rather than being a set of technologies, Web 2.0 refers to a set of practices with specific features. Specific Web 2.0 practices such as microblogging or podcasting emerged from the use of these technologies. In alignment with this view, Dohn (2010, p. 345) mentions the main features of Web 2.0 in the following way: "bottom-up" participation, collaboration, openness, distributed ownership of content, re-use of materials, unrestricted practices, and making use of digital resources.

By taking this difference of meaning in technology and practice into account it should also be emphasized that using a particular web resource such as a blog should involve a Web 2.0 practice under every condition. So, the delicate difference in meaning requires the researchers to go beyond the mere employment of a particular technology and look in depth whether the practices or values mentioned above are existent. However, I also agree with Dohn (2010) that there are several degrees of 'a Web 2.0 activity' which could certainly not be described by all of the items listed before with the exception of the requirement that it should take place on the web) which Dohn (2010) argues is a necessary condition (Dohn, 2010).

In agreement with Dohn's characterization Conole (2007) claims that due to the increased use of these technologies for socialization, three fundamental shifts are taking place: a shift from information, passive engagement and individual learners to communication, an interactive engagement, and learning practices that are situated more socially (Conole, 2007, p. 82).

The increasing number of social interactions in the fluid spaces between virtual and face-to-face environments can also be observed in the case of the CSC volunteering program. The employee volunteers not only interact during their voluntary tasks on the fieldwork; but also continue to do so through available collaboration tools. Samarawickrema (2007), studied implications of

Web 2.0 tools and came up with the finding that these tools do not only reduce the presence of the teacher but also provide individuals with a peer-feedback mechanism and an opportunity for collaborative knowledge construction.

By taking into account these scholars' views, I accept that the Web 2.0 includes several degrees in terms of its practice rather than the adoption of a range of technologies for the sake of doing so. Therefore, I prefer to focus on the degree to which the volunteering activity involving the utilization of a particular Web 2.0 tool adopts a set of collaborative practices rather than adopting a content focused approach. Yet, for the sake of clarity and brevity I shall use terms such as Web 2.0 or online collaboration to cover this wide spectrum of possible practices.

### 2.3.5 Summary for CSCL Theories

This study explores in-depth how technology is used as a medium for collaborative learning by supporting individual volunteers in terms of their communication or task-sharing on joint activities, so within that regard, theories underpinning this study are also derived from the field of CSCL. After providing main definitions for collaboration and cooperation this second part of the literature review focuses upon underpinning theories about the use of collaborative technologies including groupware.

One of the major strengths of the existing body of CSCL literature is that despite the different approaches used in CSCL studies one commonly agreed aspect of learning is its situatedness in 'peer networks' (Brown and Duguid, 2000; Hiltz, 1990). By providing individual accounts with regard to the preference for using these tools and actual usage of the online collaboration tools, this study sheds light on how this balance might be achieved within these 'peer networks' (Brown & Duguid, 2000; Hiltz, 1990) of geographically distributed online communities of volunteers.

Another strength of the literature is that when it comes to a definition of Web 2.0, it is commonly agreed that Web 2.0 refers to a set of practices rather than a set of technologies (Dohn, 2010). In alignment with this view, this study also sheds light on different aspects of Web 2.0 activities such as collaboration, openness, distributed ownership of content, re-use of materials, and making use of digital resources.

Still another strength of the literature is that despite a variety of definitions with regard to collaboration and cooperation, there is a common agreement that when it comes to collaboration individuals work with the goal of accomplishing a common task. Based on this shared understanding book identified

28  *Some Theories about Online Collaboration*

the online collaboration tools to be explored in-depth in terms of how they are being used by knowledge workers and provide support for the unplanned learning events of knowledge workers.

Within the body of literature, a framework that has been commonly agreed upon in terms of making a comparison and contrast among these studies about collaborative learning does not exist yet. This study elaborates on a framework for effective collaboration by including all aspects of collaboration such as the nature of the collaborative task, the number of collaborators, the relationship among collaborators, the underlying motivation for collaboration, the context of collaboration, the means of collaboration – whether it is computer-mediated or not – as well as the duration of collaboration. Furthermore, by providing examples with regard to the actual use of groupware (Lotus Notes) this study also contributes to an understanding about the extent to which effective online collaboration might occur due to the integration of these tools into workplace learning practices in addition to the Web 2.0 tools.

## 2.4 Conceptual Framework

Book purports to explore the usage of online collaboration tools in supporting knowledge workers for the practice of employee volunteering. Based on the literature review about the collaborative learning tools provided on the prior section, online collaboration tools refer to the web-based technologies such as popular Web 2.0 applications like blogs, wikis and traditional web-based applications such as instant messenger, online chats and messaging used by several individuals with the aim of accomplishing a common task.

The study essentially examines the use of various online collaboration tools during the implementation of a volunteering program. The emphasis in this study will be not only on the individual elements of technology or individuals, but in particular on their reciprocal interplay. The definition suggested previously emphasizes the formal and informal means through which individuals interplay as an essential aspect of the process of collaboration.

The study tries to explore three major fields: features of effective online collaborators, assumptions and motives, and approaches for effective online collaboration. As a result, the conceptual framework maps the following high level categories against the following domineering aspects (identity, relationships, feelings, control and abilities; Table 2.2).

- Descriptions of usage: This dimension of the conceptual framework refers to where participants describe how they use online collaboration tools throughout the CSC Program.

## 2.4 Conceptual Framework

Table 2.2 Main elements of the conceptual framework

| Themes | Sub-Themes |
|---|---|
| Exposure and approaches | Usage and Approaches |
| Exposure | Usage |
| Presumptions and motives | Feelings |
| Exposure | Skills |
| Approach | Choices |
| Enabling or inhibiting factors | Support |
| Exposure, assumptions and motives | Usage and Feelings |
| Exposure | Critical moments |

- Choices about approach: These include the reasons why participants use online collaboration tools throughout the CSC Program.
- Feelings about usage: This dimension provide information about the confidence, difficulties, concerns related to the use of online collaboration tools as experienced by employee volunteers.
- Sources of support: This dimension informs about who supports the user; and who the influential users are.
- Nature of support: This includes information about what type of support is provided.
- Evaluation of support: This relates to how effective the support has been considered by participants.

Assumption is a subjective term and hence might sound as possibly problematic, particularly when it comes to the dissemination and discussion of analysis and consequences of particular "assumptions" for individual and organizational uses. So, the meaning of "assumption" refers to the following definitions or statements:

- An argument or a claim that a person holds as valid or true;
- Its validity or truth is regardless of whether or not there is any evidence of the truth or validity of the claim;
- Assumptions are difficult to identify as they exist implicitly.

A review of the literature, led me to make the following conclusions regarding how the term "assumption" is used and understood in the field of digital learning:

- In terms of digital learning, assumptions refer to beliefs about both of the benefits (or not) of using online collaboration tools and the related skills required for their usage;
- The terms perceptions and assumptions are often used reciprocally in the field digital learning (Abbitt and Klett, 2007);

- When mentioning the assumptions about online collaboration reference is also made to the underlying attitudes and values. Attitudes involve a feeling or position concerning a person or a thing and they have a more affective aspect;
- One of the common emphasis points in digital learning research includes assumptions about self-efficacy (the confidence in ones' skill set to do something). For example, belief in one's ability to successfully use online collaboration tools. (Bica et al., 2005; Baylor et al. 2004).
- Another reason for the interest in exploring assumptions in digital learning research is their role as a supportive means to anticipate the practice, e.g., usage, acknowledgment, and utilization of e-learning. Scholars in this field mentioned that the successful use of a specific tool depends on the users' assumptions with regard to the extent of how much benefit will occur from its use (Bica et al., 2005; Baylor et al. 2004).

For book, the significance of focusing on assumptions is that the confidence or skill level of the users may not provide a complete explanation for the extent which these online collaboration tools differ in terms of their effective usage. It may partially be due to their assumptions about their skill for being an effective online collaborator as well as their assumptions about the relative advantages of using these online collaboration tools.

Due to the fact that assumptions are often not explicitly verbalized, I have recognized statements related to the assumptions of the employee volunteers if they referred somehow to the advantages (or disadvantages) of these technologies or if they mentioned about their skill set (or not) with regard to the usage of these online collaboration tools. Rather than questioning these statements I acknowledged them as statements of fact made by the participants. Regardless of the truth of these statements, what matters for me as a researcher is the degree by which the digital experiences of the volunteers with these tools appear to be influenced by their assumptions contained within them.

## 2.5 Summary

This chapter aimed to give a backdrop for positioning my study. Although the CSC Program took place outside the physical building of IBM, the IBM CSC case is solidly positioned in the tradition of workplace learning.

Although the study is restricted to only one employee volunteering program, it is probably one of the most unique studies ever done related to the practice of non-formal learning, a field wherein little research has been done until now about employee volunteering programs. By exploring problem areas and opportunities in the use of online collaboration tools I hope to make a contribution to the WPL in the above-mentioned domains.

## 2.6 Concluding Remarks

This chapter focused on main two streams of research due to the research focus: namely, workplace learning and CSCL. In order to place this discourse about online collaboration into the broader context of the volunteering practice, the chapter introduced two case relevant aspects of workplace learning including social aspects of workplace learning and the literature of CSCL. While the WPL was provided due to the social "situatedness" of learning (Winograd and Flores, 1986) embedded within the CSC Volunteering Program, the CSCL literature was provided due to the group level interactions among the CSC teams by use of the online collaboration tools.

Currently, the literature of both workplace learning and CSCL are in lack of research studies that provide an understanding of the assumptions and motives, and approaches with regard to effective online collaboration of knowledge workers.

Moreover, there has been no study undertaken about the use of online collaboration tools for the volunteering practice of employees. Also, studies including global teams are not available within the existing body of WPL. As a consequence, this study fills these existing gaps by focusing on the choices and strategies of IBM employees with regard to their use of online collaboration tools during their volunteering program.

As a majority of studies within the literature have focused either on the individual elements of technology or on individuals, this study shifts the emphasis onto their reciprocal interplay. The study tries to explore three major fields: features of effective online collaborators, assumptions and motives, and approaches for effective online collaboration. As a result, the conceptual framework maps descriptions of usage, choices about approach, feelings about usage, sources, nature and evaluation of support against the knowledge workers' identity, relationships, feelings, control, and abilities.

Based on a review of the CSCL literature, the conceptual framework includes main operational definitions regarding how the term "assumption" is used and understood. Assumptions refer to beliefs about both of the benefits (or not) of using online collaboration tools and the related skills required for their usage (Abbitt and Klett, 2007).

For this book, the significance of focusing on assumptions is that the confidence or skill level of IBM employees may not provide a complete explanation for the extent which these online collaboration tools differ in terms of their effective use. It may partially be due to their assumptions about their skill for being an effective online collaborator as well as their assumptions about the relative advantages of using these online collaboration tools. Book provides in-depth insights about the digital experiences of the volunteers with the digital collaboration tools and to which extent these experiences appear to be influenced by their assumptions contained within them.

# 3
# Researching the Online Experiences of IBM Employees

This chapter discusses the key research questions, the overall methodological approach, the design of the study and research methods and strategies as well as ethical issues, and a short description of data analysis which includes a pilot study as well.

## 3.1 Research Questions

The study aims to answer the following research questions:
- How are collaborative learning tools used for the volunteering practice of knowledge workers?
- What are the assumptions of knowledge workers about the benefits and challenges in using these tools for such a practice?

Wherever I quote the contributions of a CSC participant, I will quote them precisely without any correction of the grammar, spelling, or phrasing of the statements. The use of the symbol [...] refers to the cases in which words or sentences have been edited out.

## 3.2 The Origins of the Methods Used for Exploring the IBM CSC Experience

The principles underpinning the involvement of learners as research subjects have their origins in two relevant fields of knowledge: namely, the "Participatory Design" and "Participatory Research" (Seale et al., 2008). Among the conventional methods for exploring the user experiences with collaborative technologies are interviews, focus groups, and questionnaire surveys. Yet, there is still a lack of methods that provide the opportunity for the "user voice" to be elaborated more in-depth (Sharpe et al. 2005). There is a growing need for new methods and processes with regard to the engagement of users in research studies in a more meaningful way. When contributing to the user-centered research methods researchers could exploit participatory design and participatory research. I will describe each approach in turn.

### 3.2.1 Participatory Design

Participatory design is commonly used in the fields of Human computer interaction, computer science and engineering design. Participatory Design consists of three areas of knowledge (Dewsbury et al. 2004; Druin, 2007; Newell et al. 2007): comprehensive; co-participatory and user-specific design. The participation of users during the whole research and development process involves not only working directly with users, but also getting engaged with them in their real contexts on an ongoing basis and conducting an iterative evaluation of their work until an agreed solution is reached (Hanson et al., 2007). In other words, a collaborative partnership is established between users and designers. Despite the variety of participatory design methods they all are strongly grounded in ethnography in terms of ongoing observations of the user and their usage of tools throughout their daily lives (Davies et al., 2004). When the focus is on hearing the "volunteer voice" in relation to their experiences with online collaboration tools such an approach seems to be highly applicable due to the in-depth perspectives and the strong narrative offered.

### 3.2.2 Participatory Research

The main principle of participatory research is that it emphasizes research with people rather than research on people (Reason and Heron, 1986; French and Swain, 2004). This change from doing research on people to doing research with people (French and Swain, 2004) encourages everyone to own the results of the research by settings the goals and making the decisions together (Everitt et al., 1992). Similar to the participatory design, participatory research approaches the whole research as a collaborative partnership and therefore tries to get participants involved in the whole process starting with design and ending with evaluation. One distinguishing feature is that the participatory research even goes beyond the collaborative partnership and focuses on relationships without less or no hierarchy (Cornwall and Jewkes, 1995; Zarb, 1992) so that both the researcher and participant have equal status and power.

French and Swain (2004) mentioned that the voice with regard to reaching a decision can denote both the individual impact of the user as well as the collective impact. These scholars also asserted that another concept of voice refers to individuals telling their own stories of themselves in order for their voice to reach the collective about the differences and diversity in their lived experiences (Swain and French, 1998).

In alignment with these ideas as expressed by Swain and French (1998), volunteer voice is interpreted in book in the following two ways: firstly in having an impact upon the planning and evaluation of the research process (i.e., through the means of participatory methods) and second, in being provided with the opportunity to offer an explanation and evaluation of these experiences.

With regard to the participatory approaches in digital learning research, Silva and Breuleux (1994) mentioned that due the collaborative nature of the **Internet** calls there is a need for a participatory approach so that users have a say in the design and implementation of new tools and applications. It is believed that by doing so, many of the barriers and misunderstandings that often occur when introducing new technologies may be avoided. Silva and Breuleux (1994) claim that in order to fully discover the potential in new technologies participatory design may be utilized (Silva and Breuleux, 1994).

Steeples (2004) described an approach, influenced by participatory research methods, in which learning technologists were involved in creating multimedia representations of their practice that could be shared with others online, in support of their professional development. One of the reasons that Steeples (2004) gave for the appropriateness of participatory research as a method in her research was that it "brings distributed people together around common needs or problems". In book, I hoped to facilitate the voluntary participation of employees involved in the CSC Program with a view to coming together to explore the common needs or problems that volunteers have when using online collaboration tools to support their volunteering practice.

## 3.3 Definition of Participatory Research in the Context of CSC Program

Relying on the fields of participatory design and participatory research, I have borrowed Seale et al.'s definition and defined IBM CSC volunteers' participation as involving them not only as research subjects, but also working with them to obtain a collective analysis of the research issues and increase awareness for each of the constituencies that the results of the study represent (Seale et al., 2008). Based on the principle of "nothing about me, without me" (Nightingale, 2006; Nelson et al. 1998) this definition involves working directly with the volunteers in the evaluation of their use of online collaboration tools throughout the CSC Program; and encouraging employee volunteers to own the outcome by sharing the decisions about the use of these tools. By making use of the participatory approaches for the design and evaluation of online collaboration tools, it is aimed to provide an opportunity for the volunteers' voice to be heard. A method like this also takes into account Sharpe et al. (2005)'s call for approaches that makes individuals feel trusted. An open-ended methodology resulting from a holistic view of research should empower the individuals by allowing them to become the ones who highlight the important issues for them (Sharpe et al., 2005).

In conceptualizing the participatory nature of book, I have derived my approach from the framework offered by Fajerman and Treseder (2000) that specifies six different ways for involving participants ranging from no involvement at all to the involvement of the participant initiated on his own or based on decisions shared with the researcher. The methodology used in this study belongs to the group of "consulted and informed" as defined by Fajerman and Treseder (2000), in other words, I as a researcher designed the study while the participants' opinions are taken seriously. Needless to say, the participants are informed of the complete research process (Figure 3.1).

### 3.3.1 Overview of the Stages of Participation

Three key stages of participation regarding the inclusion of corporate employees can be described as follows:

- *Stage One* (January 2009–January 2010): Discussions regarding research questions and research methods suggested;
- *Stage Two* (March 2010–January 2011): Contribution of my own experiences of using online collaboration tools;
- *Stage Three* (January 2011–April 2011): Validation and interpretation of the results of the study and contribution to the design, content and dissemination of project deliverables and outcomes.

During the first stage of the study employee volunteers were asked about the appropriateness of the research questions proposed and the relevance of data collection methods suggested (Appendix J).

## 3.3 Definition of Participatory Research in the Context of CSC Program

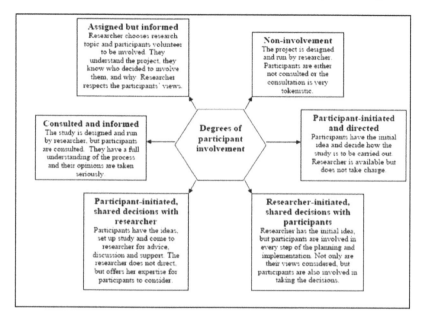

**Figure 3.1** Fajerman and Treseder (2000)'s model for levels of participant involvement.

During the second stage of the employee volunteers contributed their own experiences of making use of online collaboration tools via means of an interview and additional information (e.g. artifact) in a form and media they prefer to provide.

The third stage of the research study includes the suggestions provided by the employee volunteers who made their statements about data analysis conducted during stage two and what major conclusions needed to be made based on them. Each of these stages is to be described in the following sections.

The approaches developed did not only enable the participation of corporate volunteers, but also provided an opportunity for the participation of a wider group of stakeholders. These included:

- Involvement of project partners with regard to recruitment of employee partners;
- Involvement of senior managers of the CSC Program.

### 3.3.1.1 Involvement of project partners

Once I received the ethical approval for book from the Ethics Committee at the University of Oxford in March 2010, I contacted the managers of the CSC implementation partner in my home country Turkey, namely the UNDP office and asked for their permission to contact the volunteers on their databases. I made an agreement with them for book; they kindly accepted to email all the participants on their list on my behalf with a message about the research study and asking for employee volunteers to participate. Based on the discussions with the managers, it was agreed to send the following email message to all volunteers who completed their volunteering activity so far:

"One of our previous project managers Miss Ayse Kok aims to make a research about web-based collaboration tools, technologies and user approaches as part of her continuing studies in UK. We would be appreciate if you could contribute to this new research study as it will be a chance for you to make a difference to the way employees put these tools into practice. We need both of your specialism and experience to convey what is important for you concerning online collaboration and workplace learning through non-formal learning initiatives such as volunteering. You only need to go to the following webpage: (URL was provided) and spend a few minutes choosing some questions and ways for information-sharing that are important for you. It will be totally confidential. Information regarding her research study and related permission are also attached in the mail."

### 3.3.1.2 Involving senior managers of the CSC program

Involving a senior project manager from IBM has been an advantage in that it has helped my research study to clearly align its aims and objectives to the use strategies of the CSC volunteering teams. This has enabled me to successfully obtain support for this study from the volunteers in order to embed my recommendations into informal learning initiatives.

## 3.4 Data Collection

One of the main contributions of book is the emphasis on using the data collection tools to get participants involved in both the process and outcomes of the research, reflecting an emphasis on change, and development as identified by Pain and Francis (2003, p. 4). According to these scholars, "the extent to which participants are engaged within the research context rather than the methods and techniques employed" is one of the major characteristic of participatory research. These authors also stated that participatory approaches originated as a process by which communities can work towards change (Pain and Francis, 2003, p. 4).

## 3.4 Data Collection

This research procedure has the general aim of providing a description of the participant's personal background and (learning) context in which they made use of collaboration tools throughout their volunteering practice. Data collection has the following main sources:

- information based on the online survey,
- digital artifacts such as blogs and wikis, and
- transcripts from the interviews.

The data collection methods of book include an interview (Appendix C), an online survey (Appendix G); and review of digital artifacts (Appendix K) all of which have been utilized in both participatory design and related participatory research. A cross table in order to match the online survey and interview details was developed. Table 3.1 provides the data collection methods based on each stage.

The online survey served as a means to acquire a wider understanding of volunteers' experiences with digital artifacts, while individual case studies (online interviews) included a description of the nature of the online collaboration activities implemented by the individual and making use of the context and background. The questionnaire included both multiple choice questions and open-ended questions. The complete questionnaire can be seen on Appendix G. The questionnaire was undertaken by 11 IBM CSC participants and 1 IBM CSC Senior Program Manager although initially it was sent to 30 participants by e-mail. The survey aimed to explore the actual status of the use of these tools within the CSC program. The questions in the survey are pointed in the direction to find out about the current state of use of online collaboration tools and value seen in using these tools. For the analysis of quantitative data SPSS was used. In order to recognize the general patterns that might emerge an overall data analysis with a more descriptive basis was undertaken across the existing data set.

Foth (2004) describes how he used a combination of online surveys, interviews, and participant observation with international students in order to inform the design of on an online community. Oosteveen and van de Besselaar (2004) also used online surveys as a device to check findings from interview data in a participatory design project with the objective of design

Table 3.1 Breakdown of data collected

| Stage One-Context | Stage Two-Case Studies | |
|---|---|---|
| Survey | Interviews | Digital artifacts |
| 12 | 17 | 30 |

and development of an Internet-based smart card system to support mobile Europeans.

Mayes (2006) describes the review of an artifact along with the interview method as the "plus" aspect of an interview. During the CSC Program, an artifact might be generated by the participant and includes blogs, wikis, or any digital documents. There are examples of similar approaches in the participatory design and participatory research fields. A similar approach is advocated by Anderson et al. (2004) as a method for involving students in the design of CSCL. It has also been implemented by Riddle and Arnold (2007) in their Learning Landscape project conducted at the University of Cambridge (although in this case, focus groups were used rather than interviews).

The uniqueness of participatory research lies not in its use of online surveys and interview along with digital artifacts, as various research studies in higher education also utilized this method (Mayes, 2006; Anderson et al., 2004). The uniqueness of the research lies in the fact that the participation influenced the nature and focus of each data collection tool. As the participation of an advisory group and some employee volunteers had certainly an impact on the design of the online survey; so did the involvement of volunteers in the online survey have an impact on the design of the online interview (Appendix I).

The selection of volunteers was done in close collaboration with the Senior Level CSC Program Managers. Corporate employees who have been mostly contributing to blogs and wikis were approached to capture their experience with CSC collaboration tools. In order not make the employee volunteers feel any pressure for their contribution, candidates for participation were informed through the use of a comprehensive mailing list rather than being e-mailed individually. In case of an insufficient number of the participants based on this first e-mail, a second reminder would be again sent again through the general e-mail list. Due to the different volunteering periods throughout the CSC field work in different countries and the possibility that different participants might not be able to reach their e-mail at different times, I thought that at least one reminder should be sent. Afterwards, I decided not to send any more reminders. In case of any interest expressed by a volunteer for participation in the study, related information were sent. If no response can be received within 2 weeks I assumed that they have withdrawn from the study and did not contact them again. The Right to withdraw was also emphasized in the Information Sheet (Appendix B).

The Information Sheet (Appendix B) also addressed the issues of how to deal with anonymity and confidentiality. An informed consent form (Appendix B) was utilized for getting permission from employee volunteers to use data and disclosure of information to third parties. In summary, all information that could be identified had to be disclosed with an explicit consent.

The combination of methods provided a fertile ground for both rich empirical data and the triangulation of interpretations of the data based on the various methods and target audience. The sampling strategy was pragmatic to some extent, collaborating with the related managers to specify the relevant volunteer cohorts to target.

### 3.4.1 Description of Stage One

With the support of the non-profit implementation partners the recruitment email was sent out to all volunteers who participated in the CSC Program so far. The email provided volunteers with information about the study and initial questions.

Those volunteers who decided to participate were asked two main questions (Figure 3.2). A pilot study with regard to these questions and their way of being asked was conducted with a volunteer who gave some useful suggestions (Figure 3.3).

Volunteers' responses for the suggested research questions showed a preference for research questions 1, 2, 3, 5, and 6 and showed a weaker response for questions 4 and 7 (Table 3.2). Some of the participants made useful comments about the phrasing or content of the suggested questions:

"There is a good diversity of these questions. The ones that I have indicated feel relevant to the way that I have used these digital tools during volunteering practice. Another question that might be asked is "how do you feel digital resources could be improved to add value to your volunteering practice" or something similar as this will provide an opportunity for the expression of different points of views. "Do you think Web 2.0 supports collaboration and volunteering activities?" as a suggestion for alternative question."

The responses of the volunteers for the suggested data collection methods showed a strong inclination for the provision of web-based links to digital resources or artifacts developed by the volunteers on their own (Appendix K for more examples). These include links to online blogs, links to the existing resources (e.g., a wiki, audio file, and blog etc.) and contributing resources to the CSC website.

> *Kindly find below the questions to be answered in Stage Two of my research study.*
>
> Please feel free to check the ones to indicate that you think is important!
>
> Then please provide any comments, revisions or additional questions in the space shown below.
>
> 1. How do you put technology into use for the purpose of your volunteering activity during the CSC Program?
>
> 2. How does technology affect your volunteering activities during the CSC Program?
>
> 3. How do you feel about making use of technology to help you collaborate with other volunteers?
>
> 4. How do you use popular Web 2.0 tools such as Facebook or Twitter for collaboration and do you think that they make any contribution to the process of learning?
>
> 5. How are you supported with regard to the use of these tools by your colleagues?
>
> 6. How do you feel about the support you have received so far from your colleagues?

3.4 Data Collection    43

> 7.    Are there specific events that led to a change in the way you used these tools for your volunteering practice?
>
> *These are some of the ways that CSC participants will hopefully share their own experiences about the online collaboration tools used by them.*
>
> Please feel free to check the ones to indicate that you think is important!
>
> Then provide any comments and other ways volunteers can contribute their thoughts in the text box below the suggestions!
>
> a.    Links to your blog
> b.    Links to your online knowledge repository (e.g., a wiki, PowerPoint presentation, web page etc.)
> c.    Contributing materials to the CSC website.

**Figure 3.2**  Questions presented to Phase One participants via e-mail.

I also discussed at length via e-mail more issues that might arise with a volunteer who had been involved in the piloting. It was found that many needed further explanations and were then happy to offer alternatives to the questions on the website. These alternatives and the responses to the online survey formed the basis for the final questions included in the phase two interview. The responses that volunteers provided in Stage One also offered practical information that supported the revision of the questions as a preparation for the interviews in Stage Two (Appendix I for more detail).

44  *Researching the Online Experiences of IBM Employees*

> I think that it would be much easier if you split the questions up into categories, so they are a bit more logical. Personally, I think there are three distinct categories:
>
> 1. How they use digital tools (facts)
>
> 2. Experience of these tools (feelings)
>
> 3. How this facilitates collaboration (does it actually help?)
>
> I'd be really interested to find out about the question regarding Web 2.0, I know some employee volunteers who use Facebook, Twitter etc. in order for them not to feel so isolated. Web 2.0 is a really hot topic at the moment. I'd be interested more about the volunteer experience and how online collaboration plays apart in this.

**Figure 3.3** Example of the quality and quantity of feedback given by a participant in the pilot study.

### 3.4.2 Description of Stage Two

The focus of the second stage was on the actual individual experiences. As 17 participants from the IBM CSC Program participated in this stage in terms of the personal accounts I will refer to 17 participants. Of these participants, 6 were female and 11 were male. The countries that the CSC participants were sent to and that they originate from differ. The majority of participants were aged 30 or over. Most participants possessed more than 5 years of experience in IBM and held higher level positions. Although the participants' interaction with the tools differs, all of them were using these tools along with face-to-face meetings with throughout the CSC Program.

Stage Two involved participants telling about their own experiences of making use of online collaboration tools through an interview conducted via **Skype** and additional information provided (e.g., digital artifacts such as blogs and wiki) in a form and media of participants' preference (Appendix K). In addition to the involvement of Stage Two participants based on the sample of interested Stage One participants, participants were involved in Stage Two

Table 3.2 Frequency of responses (tick marks) to each proposed research question

| Sub-Themes | Research Questions Suggested | Number of Checkmarks Received |
|---|---|---|
| Choices and use | 1. How do you put technology into use for the purpose of your volunteering activity during the CSC Program? | 12 |
| Use | 2. How does technology affect your volunteering activities during the CSC Program? | 11 |
| Feelings | 3. How do you feel about making use of technology to help you collaborate with other volunteers? | 10 |
| Choices | 4. How do you use popular Web 2.0 tools such as Facebook, Twitter and LinkedIn for collaboration and do you think that they contribute to your learning? | 4 |
| Support | 5. How are you supported with regard to the use of these tools by your colleagues? | 10 |
| Feelings and Use | 6. How do you feel about the support you have received so far from your colleagues? | 9 |
| Critical Moments | 7. Are there specific events that led to a change in the way you used these tools for your volunteering practice? | 5 |

through a combination of purposive sampling and the snow-ball technique. Purposive sampling and snowball sampling was adopted in order to try to ensure that a range of different background were represented in the 17 case studies. According to Sharpe et al. (2005) it was crucial important to explore how the use of web-based tools influenced the learning experience of participants. So, these scholars suggested that user experience studies should select the sample participants driven by a purpose. Purposive sampling involved targeting employees who were contributing to blogs and wikis frequently. It was foreseen that by targeting these individuals participants who might provide deep insights into the underlying reasons for the use of online collaboration tools might be involved. Snowball sampling involved asking participants if they could suggest of a friend who might be interested in taking part in the study and if so, if they could convey the related information about the study to them. Of the original 17 volunteers who participated in Stage Two, five got involved in the group as a result of being informed by another

CSC team member and the remainder joined as a result of the e-mails sent out by as described in earlier.

In Stage One, volunteer responses to the suggested data collection methods revealed a strong inclination for the provision of related project links and digital artifacts that they had created on their own. By making use of participatory methodologies, it appeared that the emphasis with regard to these artifacts would be the approaches adopted by volunteers for their use of digital tools to support their volunteering practice. The media that volunteers chose to represent their strategies ranged from online blogs, links to the existing resources (e.g., a wiki, PowerPoint presentation, and web page etc.) to additional digital resources on the CSC website.

Seventeen interviews are conducted within over a 6-month period in total. Similar to the traditional interview, the online interview started with an introduction, including a brief overview of the study and agreement for confidentiality, "warm up" questions, more specific questions with a focus on the core themes, and final questions with open-ended styles like "Would you like to make any final comments?" During the interview, volunteers were asked about the different kinds of digital tools they used for online collaboration during the CSC Program. Based on this information, a series of semi-structured questions guided the conversations. I did not attempt to explain what kind of answers is sought, or to raise the issues which might lead to the desired responses so that the volunteers did not feel under any influence. Through the content analysis technique online recordings of the interviews are transcribed as case descriptions for interpretations. In order to maximize my memory recall I put an effort to complete each interview transcription within the same day of the interview.

Table 3.3 provides a general picture in terms of the alignment between research questions, interview questions and project themes. With the approval of volunteers, all interviews were recorded and transcribed. An interview lasted around 40 minutes on average. Due to the participatory nature of the study the interview along with data collection period typically required two to three online meetings with volunteers in order to complete the interview; determine the strategy or strategies used by the volunteer with regard to the use of collaboration tools throughout the CSC Program.

The accounts offered in the next sections provide information with regard to the usage of the online collaboration tools throughout the CSC Program and explores the approaches implemented when putting these tools to facilitate their participation in the CSC Program.

**Table 3.3** Alignment between the research questions and interview questions

| Research Question | Interview Question |
|---|---|
| • How are collaborative learning tools used for the volunteering practice of knowledge workers? | 1. How does your organization make an effort to contribute to the usage of online collaboration tools during the CSC Program? |
| • What are the assumptions of knowledge workers about the benefits and challenges in using these tools for such a practice? | 2. What are the main factors that allow/limit your organization to facilitate the use of the use of online collaboration tools within this context? |
| • How are collaborative learning tools used for the volunteering practice of knowledge workers? | 3. How many times in a day do you use any of the online collaboration tools to exchange information with your colleagues and other related individuals involved in this CSC Program? Please give me some examples of what you use and how you use it. |
| • What are the assumptions of knowledge workers about the benefits and challenges in using these tools for such a practice? | 4. Which tool does give you the best opportunity to provide knowledge sharing opportunities with your colleagues? |
| • What are the assumptions of knowledge workers about the benefits and challenges in using these tools for such a practice? | 5. Are there any downsides to using online collaboration tools for professional knowledge-building and -sharing? For example? |
| • What are the assumptions of knowledge workers about the benefits and challenges in using these tools for such a practice? | 6. Do you think using technology – specifically for collaboration in this CSC Program can be improved? Please give specific examples. |
| • What are the assumptions of knowledge workers about the benefits and challenges in using these tools for such a practice? | 7. What are your key concerns of the use of online collaboration tools? |
| • How are collaborative learning tools used for the volunteering practice of knowledge workers? | 8. What are the factors that can contribute to your engagement with online collaboration tools? |
| • What are the assumptions of knowledge workers about the benefits and challenges in using these tools for such a practice? | 9. What are the benefits that you hope to gain in return from your contributions to the exchange of idea via the use of online collaboration tools? |
| • How are collaborative learning tools used for the volunteering practice of knowledge workers? | 10. Is there anything else about your use of online collaboration tools that I could have asked you? Or anything else you would like to add? |

The way the results are presented in the next section, has been affected by several main factors. To begin with, I would like the volunteers to provide with the opportunity to tell their stories (Munford et al. 2008) despite my acknowledgement that as a researcher I also acted as a mediator for telling these stories. Second, I think that as a researcher I should keep a distance between myself and the study that presents information into the form of a "case record", as this would categorize some information as privileged based on its degree of usefulness for professionals(Gillman et al., 1997); in other words the IBM employees. My intention is to convey knowledge that provides a contextual conceptualization with regard to the experience of using online collaboration tools throughout the CSC Program. While the individual experience of the volunteers will be conveyed through the use of personal accounts (Appendix F); the collective experience will be conveyed through the extraction and presentation of quotes from across the interviews in order to highlight the themes and refer to the commonalities and diversity among them. In addition to the quotes, whenever appropriate, related research studies will also be referred to when discussing the CSC participants' experiences of the use of online collaboration tools. To avoid the reader from having a decontextualized feeling of the experience of the volunteers just by getting exposed to a list of extracted quotes, I made an effort to convey the personal accounts of all of the employees. Rather than selecting out just some of accounts to present here and to disempower some participants while privileging others I argue that the presentation of all personal accounts is more helpful.

Table 3.4 provides an outline for the suggested approach for coding the interviews. Yet, when the coding frame was put to use within NVivo, I realized that through my preliminary analysis the data seemed to be more relevant to be coded by NVivoin comparison to other areas. This made me to revise my coding framework. Reasons for the simplification of the framework included the fact that descriptions of how volunteers make use of the tools to support their participation in the CSC Program and the strategies they adopt were better shown through the individual accounts.

### 3.4.3 Description of Stage Three

Stage Three of the study focused on:

- Involving participants in validating and interpreting the outcomes of the study;

**Table 3.4** Comparison of the frameworks for preliminary and final interview coding.

| Preliminary Framework for Coding | Final Coding Framework |
|---|---|
| Descriptions of Usage (i.e., where participants describe how they use online collaboration tools throughout the CSC Program) Feelings About Usage (confidence, difficulties, and concerns) | Feelings and Assumptions About the Use of Online Collaboration Tools |
| Support—Sources (who provides the support; individuals having an influence) Support—Nature (what kind of support) Support—Evaluation (the extent to which the support perceived to be as useful or effective) | Evaluation of Support |
| Choices About Approach: Reasons why participants use online collaboration tools throughout the CSC Program | Decisions for Using Online Collaboration Tools |

- Enabling volunteers to make a contribution to the design, development and dissemination of deliverables and results.

The first aspect of participation involved transcript validation. The second aspect of participation involved collaboration over the design of an improved online collaboration platform for the CSC Program.

### 3.4.3.1 Transcript validation

After the interview transcripts had been typed they were sent to volunteers by e-mail for correction and additions. In general, this was carried out by the participant making use of comments or corrections in Microsoft Word. The corrections were mainly due to poor transcribing or further explanations around the topic discussed; in only one case did a volunteer ask for a few sentences to be removed from the transcript. Three participants sent back corrected documents, the rest made comments in e-mails. None of the participants expressed concern that the findings would falsely represent the experience of individuals. In this way, it was ensured that the participation was inclusive in nature, where all participants have a chance to contribute and are under no pressure to acquiesce and agree with the majority view.

Giving interviewees the chance to validate their interview transcripts is not a technique that is unique to participatory research; it is quite common in qualitative social sciences that use interviews or focus groups. However, it is an essential element of participatory approach in relation to ensuring that the "voice" of the participant is the voice that the participant would want to be heard (Sharpe et al., 2005).

### 3.4.3.2 Design of CSC collaboration platform

Once the online interview came to an end, I sent the participants three different mock-ups of what the interface to the new online collaboration platform would look like and asked for their views and preferences via e-mail (Appendix L). I followed this up with an email to each participant giving them a URL to an online submission form which contained screen shots of various ideas for browsing through the strategies provided by participants. Each screenshot option had a box that participants could click if they preferred that option. Participants could also add comments on their choice at the bottom of the form. Submissions were anonymous.

Around 15 participants mentioned their opinion about the design (look and feel) of the new collaboration platform. The volunteers also added commentaries and explanations regarding their responses (Appendix H). The responses are detailed and give clear explanations for preferences. On the whole, participants chose the cleanest look. Also, two participants participated in the technical design of the new platform to a significant extent.

## 3.5 Data Analysis

To conduct the quantitative data analysis SPSS was used, while for qualitative analysis Excel was used by separating content into appropriate sections and manipulating it. Open comments provided about the answers were put into an additional column in the Excel file. In order to see whether some general patterns emerge an overall descriptive analysis was conducted based on the available dataset. A further analysis of these patterns showed whether there were differences among the volunteers. Based on the emerging patterns, I coded the qualitative data and ranked the results or directly quoted to support the quantitative findings.

After data collection at the level of individual participants, I tried to put each case study into analysis individually followed by an overarching study across the cases (study of cases). The main purpose of the qualitative data analysis was to extract and abstract from the complex data any evidence with regard to the activities and experiences with online collaboration tools to convey responses to the research questions. I transcribe relevant extracts from the interviews to supplement the results of the survey. I used this analysis to convey more detailed information about the approaches that the participants put into use and in which ways the tools had an impact on both their approach to collaboration and their knowledge-sharing activities.

As a strategy for the validation of the results, triangulation is used to further increase scope, depth, and consistency in methodological proceedings (Patton, 1990). The triangulation in this study is based on three perspectives for data interpretation, namely online interviews, questionnaires and review of digital artifacts that are verified by also the IBM volunteers (Patton, 1980). During the analysis phase, transcription was undertaken mostly by the use of the conventional approach of recording, pausing and typing. Transcripts once decoded were sent to volunteers for revisions by e-mail. These transcripts have established the ground to recognize the issues and construct artifacts based on strategies. For further analysis, all verbatim transcripts of the online interviews with the interviewees were imported into NVivo. Table 3.5 provides an overview of the alignment of suggested coding categories with research questions and interview questions. Digital artifacts such as entries into the CSC Program wiki, blog, or Lotus Notes tools served as supporting sources. The themes and the categories to which they belong have been changed in case of any differences until a common agreement has been reached among the participants.

The broad interpretive framework for the study involved the exploration of specific cases and individual opinions about the way the volunteers used digital tools to support their collaboration throughout the CSC Program. By means of ethnographic approaches the kinds of technologies and strategies used throughout the CSC Program were identified together with the volunteers' experiences.

## 3.6 The Challenges of Using Participatory Approaches

A few main challenges for participatory research have been specified throughout the research literature. According to Ward and Trigler (2001), one of the main challenges is to feel confident that participation really has an influence on the extent to which the research study can respond to the questions of individuals being subject to the inquiry. Another challenge relates to ensure that results be not mere stories reconstructed or validations of research conducted (Duckett and Pratt, 2007). The extent to which participants are capable of contributing to the analysis and interpretation of data, also needs to be considered (Richardson, 2000). The experience of conducting the research study revealed five main challenges with regard to the use of participatory

Table 3.5 Overview of the alignment of the suggested interview coding categories with research questions and interview questions

| Research Questions | Mapping with Interview Questions | Mapping with an Interview Coding Framework |
|---|---|---|
| How are collaborative learning tools used for the volunteering practice of knowledge workers? | 1. How does your organization make an effort to contribute to the usage of online collaboration tools during the CSC Program?<br>3. How many times in a day do you use any of the online collaboration tools to exchange information with your colleagues and other related individuals involved in this CSC Program? Please give me some examples of what you use and how you use it.<br>8. What are the factors that can contribute to your engagement with online collaboration tools?<br>10. Is there anything else about your use of online collaboration tools that I could have asked you? Or anything else you would like to add? | Descriptions of Usage (i.e., where participants describe how they use online collaboration tools throughout the CSC Program)<br>Choices About Approach: Reasons why participants use online collaboration tools throughout the CSC Program |
| What are the assumptions of knowledge workers about the benefits and challenges in using these tools for such a practice? | 2. What are the main factors that allow/limit your organization to facilitate the use of online collaboration tools within this context?<br>4. What are the main factors that allow/limit your organization to facilitate the use of online collaboration tools within this context? | |

5. Are there any downsides to using online collaboration tools for professional knowledge-building and sharing? For example?

6. Do you think using technology – specifically for collaboration in this CSC Program can be improved? Please give specific examples.

7. What are your key concerns of the use of online collaboration tools in relation to knowledge-sharing?

8. What are the factors that can contribute to your engagement with online collaboration tools?

Feelings About Usage (confidence, difficulties, and concerns)

Support—Sources (who provides the support; influential people)
Support—Nature (what kind of support)
Support—Evaluation (how useful or effective was the support perceived to be)

methods: informed participation; valued participation; non-hierarchical participation; empowered participation; and transformative participation.

## 3.7 Ethical Considerations

An ethical approval for the research study was formally acquired through the submission of an application to Ethics Committee (Appendix D). To ensure that the research is subject to appropriate ethical scrutiny, the CUREC form has been completed and consent has been received (Appendix D). Ethical issues that might occur throughout the research study, in consultation with all participants involved have been carried out using the checklist and CUREC's other documentation.

Two main issues with regard to the ethical approval had been determined that should be addressed in more detail:
- Issues relating to gaining access to the employee volunteers;
- Issues relating to anonymity and confidentiality.

### 3.7.1 Gaining Access to Participants

There is one important key gate-keeper in this project, who has the ability to provide indirect access to volunteers across whole IBM. He is the Senior Level CSC Program Manager. He had the contact list of all the volunteers who completed their registration for the CSC Program. Before contacting him, he has been informed by my director about my research study and my background as a researcher. Then, I have been provided with his contact information by my director. After having introduced myself I gave him information about my research study (e.g. a document that can be sent as an email attachment) and asked him kindly to distribute it via this email list. He kindly agreed to support me throughout my research study. I also informed him about the documents that will require his signature based on the ethical guidelines within my university.

To ensure that volunteers have a detailed understanding about the nature of the research an Information Sheet (Appendix B) has been prepared. To ensure that volunteers make their own decision about taking part an Informed Consent Form has been used (Appendix A). Through Informed Consent Form (Appendix A) participants have been told that they can withdraw freely at any time.

Once I received the ethical approval for book from the Ethics Committee at the University of Oxford in March 2010, I contacted the managers of the CSC

## 3.7 Ethical Considerations

implementation partner in my home country Turkey, namely the UNDP office and asked for their permission to contact the volunteers on their databases. I made an agreement with them for book; they kindly accepted to email all the participants on their list on my behalf with a message about the research study and asking for employee volunteers to participate.

### 3.7.2 Anonymity and Confidentiality

Anonymity and confidentiality were crucial especially with regard to the ethical procedures concerning the digital artifacts. In terms of anonymity and confidentiality, participants were guaranteed full anonymity. This has also been emphasized during the initial discussions throughout the involvement stage of the Senior Level CSC Manager of IBM. Once his confirmation for this study has been received, I also ensured throughout online discussions with IBM employees that that neither their full name nor any of their online identity including their social media profile will be disclosed in my research study. In addition, any type of digital artifact that they created such as web sites or blogs will also not publicize their identity. In other words, I will be the only one accessing their personal digital spaces such as internal blogs within IBM Lotus Notes. This has been agreed upon mutual trust. The Information Sheet (Appendix B) addresses the issues of how to deal with anonymity and confidentiality. The informed consent form (Appendix A) also includes issues with regard to conveying information to third parties and obtaining approval from volunteers regarding data usage. In summary, no specific information which could be identified by others was to be disclosed without explicit consent.

# 4

# Results

This section elaborates on the volunteer voices in a more detailed manner by providing information about the in-depth case studies. In the following sections, I present contextual background information about the tools before proceeding with the analytical themes that emerged from the data collection.

## 4.1 Contextual Background Information about the Tools Used by CSC Participants

IBM is one of the oldest companies in terms of its adaptation and use of online communities. Topic-related online communities have been in use for more than 10 years and provide an online space for professionals according to their area of expertise with various communication facilities such as discussion forums, blogs, wikis, **instant messaging**, and social media platforms. In addition to the online public spaces with full access, in other words sites which can be viewed by all IBM employees, there are also areas which require registration for access, those which require the confirmation of a site administrator, as well as private online communities which members can only join upon private invitation. These online spaces enable employees with shared interests to establish cross-departmental communication to provide a basis for the transfer of their knowledge and experiences. All of these tools are available for use by participants of the CSC Program and there is no mandatory requirement for IBM employees for using these tools, yet its use is strongly recommended by the management and the tools have become also an important part of their workflow ad daily communication within IBM. When making use of these tools IBM employees need to adhere to internal social computing guidelines. Furthermore, it is important that tools are easy to use for the employees. For instance, a blog entry can directly be published from any computer by an employee by making use of an office application or the web browser. No significant hardware requirements exist so that employees find the use of these tools easy; the only limitations that exist for these tools relate to the file sizes.

58  *Results*

The main collaboration tools used throughout the CSC Program are based on Lotus Notes, Domino, the IBM intranet (including blogs, wikis, and discussion forums), and e-mail. In addition to this, the internal **VLE** called Edvisor also offers various learning modules to prepare the participants of the CSC Program for their fieldwork.

## 4.2 Participants' Use of the Online Collaboration Tools

In order to understand about the usage of the tool by IBM volunteers for the purpose of exchanging their experiences and knowledge with others, it is important to understand which tools of the ones available that are used. This section provides some information about the participants' usage of the collaboration tools made available to them throughout the CSC Program. By doing so, my aim is not to make any interpretations with regard to the generalizability to some other group of volunteers, rather it is to provide furthering contextualization and description.

It is clear that some tools are used more readily than others as represented in the following graph (Figure 4.1).

Identifying which tool the participants feel comfortable with leads to their motivation to use the tools more readily to share their own experience, knowledge, and best practices in general. All 12 participants stated they were motivated with five indicating they were very motivated to share. However,

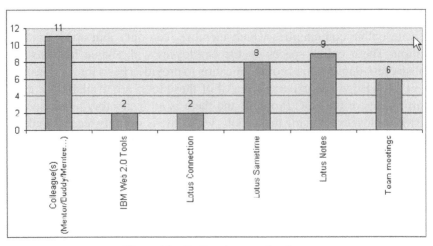

**Figure 4.1**  Preferred usage of tools.

## 4.2 Participants' Use of the Online Collaboration Tools

sharing does not always lead to collaboration. In this instance though, the individual propensity to share knowledge for collaboration is also important. Most of the participants thought sharing of topics and knowledge is always important. The CSC team members know that they can learn from other team members, are motivated to share experience, knowledge and best practice which can facilitate the ground for encouraging the usage of online collaboration tools from a motivation's point of view.

Furthermore, support in terms of being encouraged to share information with the whole team and all team members, was important but as some of the responses indicate that they are not supported or not enough supported to share an appropriate level of information with the team; it shows that there is room for improvement.

Question 8 on the other side is used to reflect on the team's point of view of being supported enough to receive enough information from other CSC team members as support is also an important aspect of collaboration. Most of the participants feel this situation is improvable, only one respondent feels it is acceptable and two feel that they received sufficient support to acquire enough information from all team members (Figure 4.2).

The next question in this context is question 9 focused on the part of finding out what are main prohibitions for sharing their experiences within the team. The results are reflected in the following diagram (Figure 4.3).

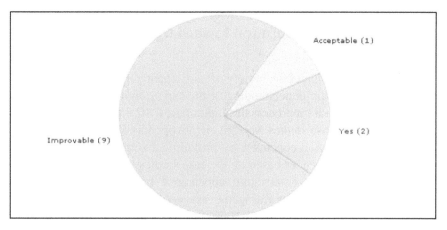

**Figure 4.2** Q8: Do you think you are able to obtain sufficient information from all other CSC volunteers?

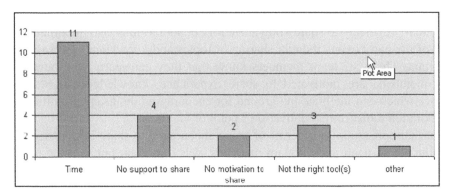

**Figure 4.3** Q9: What is/are the prohibition/s in case you think you are not able to provide, share and acquire best practices, knowledge and leverage the experience of CSC volunteers?

Looking at the results it shows that especially the factor time is one of the main prohibitions of sharing CSC related knowledge, best practices and experiences within the team. The answers state on the other hand that "no support to share" is existing, they have no motivation to share and that the right tools are not existing.

These technologies and methods for collaboration depicted in the survey correspond well to the digital tools mostly mentioned in in-depth study interviews; these will be explained in more detail in the following sections.

## 4.3 From Online Collaboration toward Distributed Cognition

The following sections depict the variety of participant perspectives in detail to provide a more holistic understanding of the use of digital tools to aid all aspects of their collaboration process throughout the CSC Program. Data with regard to all of the 17 case studies were given in Appendix F (summary of the 17 interviews). The section concludes by providing a summary of the main commonalities and differences based on the case studies.

The main picture that comes into appearance based on participants' accounts is that volunteers are fitting the usage of these technologies for their own personal, individual needs – mixing usage of general ICT tools and resources, with company-specific tools and resources.

It would seem that new forms of collaboration which come into emergence due to the interaction of team members with new Web 2.0 technologies imply

## 4.3 From Online Collaboration toward Distributed Cognition

a transformation toward Salomon's (1992) notion of 'distributed cognition' and shared enterprise with tools. These digital tools help to display ideas and are a quick way to collect and collaborate information. Informal and formal distributed cognition are apparent throughout the CSC project. For example, Sametime, which was one of the tools recommended by IBM was utilized amongst the volunteers to do and plan group tasks in collaboration. Sametime which is a free application to use does not enable individuals to communicate effectively with others, but also allows one to share files across it, and provides the ability to work on group projects, and to have a video conference. Sametime was also utilized for more informal discussions with colleagues if a subject is not understood. Sametime, like e-mail, was considered more efficient than other forms of communication as it consumed less time in comparison to others (Table 4.1). Tools such as Sametime provide a supportive mechanism by which volunteers distributed in space and time over the length of a project can interact and build knowledge collaboratively.

**Table 4.1** Use of collaborative tools used among CSC participants

| Pedagogy Approaches | Supporting tools and outputs | | Interactions |
|---|---|---|---|
| | Tools | Output | Use Types |
| Collaborative | VLE (Edvisor) | Presentation | More group interactions, group perspectives – social dialogue |
| | Wikis | Report | |
| | | Review | Increased level of social visibility, feelings with regard to higher level of confidence and belonging to a community |
| Dialogic | Web Conferencing | Artifact | More formal written interaction, task-based |
| | Discussion board (**Lotus Connection**) | | More individual contributions, personal perspectives |
| | Instant messaging (Sametime) | | Lower social presence |
| Field trip | Blogs | Artifact | More informal verbal interaction, with social activity |
| | | Review | |
| Project-based | Email (Lotus Notes) | Artifact | More formal written interaction, task-based |
| | Lotus Connection | Presentation | More individual contributions, personal perspectives |
| | | Report | Lower social presence |
| | | Review | |

A summary of the components related to the volunteering tasks is provided in Table 4.1. It has been adapted from the taxonomy developed by Conole and Fill (2005) which includes the related context of the activity taking place (learning outcomes, etc.) and the tasks undertaken. It should be taken into account that Table 4.1 should not be interpreted as if the boundaries in relation to the use types of the tools used among CSC participants are clearly drawn. For instance, while a dialogic approach might utilize mostly discussion forums and blogs, it can also involve wikis. The Table 4.2 depicts the tools most commonly deployed, depending on the particular approach.

### 4.3.1 Participant Approaches

In line with the approaches above the CSC participants devise and adopt a variety of approaches when using technology to support their volunteering process. The most common types of approach adopted by participants seem to be related to sharing of experiences, knowledge, and best practice which can establish the ground for encouraging the use of online collaboration tools. These are outlined in Table 4.2.

It should be mentioned that these themes will be explained in Section 4.6.4. These approaches offered at the beginning often seemingly related to the standard use of Web 2.0 tools given the popularity of these tools. However, it often transpired that the volunteers developed a range of refined and customized approaches based on the social affordances of the tools such as being able to create co-presence or extend their individual space. It may be an alternative tool such as Twitter or Facebook instead of the classic Lotus Notes community. It may be related to collaborate via means of a wiki due to the requirement of file sharing or it could be the fact that a mixed usage of different types of tools was required to facilitate collaboration. Obviously, these choices were in place due to personal reasons, easiness in terms of usage and the extent to which they were appropriate for the situation at hand.

## 4.4 Decisions for Using Online Collaboration Tools

When talking about the adaptation of activities and technologies to suit one's own circumstances Creanor et al. (2006) used the term "sophisticated awareness". Given the scope and novelty of approaches used by the CSC participants, it would seem the volunteers go beyond developing an awareness to suit their own circumstances and display a sophisticated way of collaboration to make the best out of their circumstances. By constantly reflecting their

**Table 4.2** Overview of the types of approaches used by CSC volunteers

| Volunteers' Approaches | Examples |
|---|---|
| Having co-presence/Sharing experience, knowledge, and best practice | Using Instant messaging; participating in discussion forums; or uploading videos or photos onto the Internet |
| Meeting new colleagues and experience parts of the world | Having discussions through Lotus Notes communities and tracking the experience of the participants |
| Navigating through information, find personal routes and pathways | Using internal Lotus Notes platform |
| Increasing one's knowledge on CSC Program | Using Edvisor especially before leaving for fieldwork |
| Reflecting on one's experiences | Blogging |
| Accessing, creating, sharing and continually improving ideas | Participating in exchange of ideas via blogs and wikis |
| Participating in networks of distributed volunteers engaging in activities | Using popular Web 2.0 tools such as Ning and FaceBook |
| Facilitating ongoing communication, dialogue and shared activity | Creating digital artifacts |
| Supporting one's learning process | Attending online trainings on culture, security and literature reading on social responsibility projects online |
| Receiving informal support | Using Skype or MSN to communicate with others |
| Aiming toward a common goal of knowledge creation | Participating in exchange of ideas via Lotus Notes communities, e-mail, and wiki |
| Participating in a team evolution process | Observing others' online behavioral pattern on discussion forums or the wiki |
| Supporting online communities and relationships between people | Participating in Lotus Notes communities |
| Having a more authentic collaboration through the creation of digital artifacts | Posting mainly on blogs or contributing to wikis |
| Recombining the information shared by others to create new concepts, ideas, and services | Utilizing Web 2.0 tools (mostly blogs and wikis) |

personal and collective endeavors situated within the context of volunteering onto the digital realm they were engaged in digital activities that embody ongoing collaborative processes. The mere existence of collaboration tools is not enough on its own without having an awareness about the usage of these tools for knowledge discovery. So, collaboration involves multidimensional interactions among all of the tools, context and users via distinctive usage patterns.

This participatory research takes a step forward toward deciphering factors that may empower employees in their collaboration process within the context of volunteering. As the study advocates the voice of the employee volunteer the factors were driven by the voice of the employees rather than driven by higher level management. A number of crucial factors were determined for decision-making and the expression of preferences about whether and how to use digital tools to aid the process of collaboration: personal factors; affordances of technologies, and properties of technologies.

### 4.4.1 Personal Factors that Influence Decisions

Personal interaction was still the preferred way of sharing practice. Yet, for those who prefer to use online collaboration tools, there are some personal factors related to the use of online collaboration tools. Two personal factors that seemed to have an influence on participants' decisions about the use of digital tools are:

- A tendency to feel part of a shared endeavor (eventedness);
- A feeling of being present with other people (co-presence).

For some participants, central to staying connected throughout the CSC Program was the provision of equipment such as **USB** or a notebook that enabled wireless connection to the Internet whatever the location as opposed to being desk-bound. For example,

> "The key factors that enable IBM to facilitate us the use of online collaboration are the equipments that they provide to us, like our thinkpad, usb, AT&T. Those 3 factors allow us be connected in everywhere at any time." CSC Participant [11]

For others, the online applications provided played a more crucial role for staying connected throughout the volunteering program and supported their co-presence. Co-presence refers to both a mode and a sense of being with others (Zhao, 2001). For instance, Participant One mentioned that the use of the free **social network** Ning was also essential to the success of their teamwork. This volunteer chose Ning over an internal Lotus Connections Community so that other stakeholders outside of IBM could easily collaborate with them as well. Ning was mainly used to share summaries of group meetings, post assignments (e.g., information about the host country), share articles, and get logistical information.

## 4.4 Decisions for Using Online Collaboration Tools

Although in the CSC Program, there weren't any mandatory requirements about being engaged in the online collaboration activities the volunteers mentioned that by perceiving the social affordances of these tools they felt more engaged:

"[…]During our assignment, the facilitator and the CSC management team encouraged us to use Facebook and Twitter to share our experiences. Our facilitator insisted that we write on the IBM CSC blog about our daily life and our assignment. We as a team noted down our experiences on the CSC Malaysia Team 4 blog." CSC Participant [3]

These statements again reveal that individuals prefer to communicate their experiences through the available tools which increase their feeling of eventedness for volunteers. As Collis and Moonen (2001) put it, engagement and ease of use affect the feeling of eventedness for volunteers.

It is clear from the statements that the participants put an emphasis on being connected through the CSC Program whether through the means of web-based tools or voice suit facilities:

"I feel more connected to my team members, hopefully for the long-term. […] Stimulating ideas, new insights and perspectives, synergies and something big being created from someone planting a small seed, engagement with others, building connections …." CSC Participant [10]

As Schrage (1990) states, the goal of a collaborative technology is to create value through human interaction rather than only providing information. The volunteers' preferences were in alignment with Schrage's thinking.

One of the most frequently mentioned affordances of digital tools referred to by CSC volunteers in terms of their reasons for the use of these technologies to contribute to their process of collaboration throughout the CSC Programme, was the 'cathartic' nature—as one participant put it—it offered them in terms of being able to reflect on their experiences and to learn about different point of views. Also participants were able to reflect on the collaborative process:

"We are also asked to chronicle our experience using new media tools such as blogs, video and image collages […]. Also, I found writing a blog while I was on my assignment to be very 'cathartic'–it enabled me to pause and reflect on what was happening to me and how I wanted to be with what was coming next." CSC Participant [1]

66  Results

The CSC volunteers do not only feel comfortable with the use of these technologies, but also view it as indispensable for their volunteering process. CSC participants have developed an awareness of the advantages and disadvantages of the use of different tools and don't utilize these tools just for the sake of it—a meaningful aim and an obvious personal gain, namely eventedness and co-presence should exist. They have an expectation of being able to stay connected to others and see this as integral to their collaboration. Rather than viewing the digital tools as something special they view it holistically and consider them as just another important aspect of their volunteering process.

### 4.4.2 Affordances of Technologies that Influence Decisions for Collaboration

In this section, I probe the influence of affordance in nurturing and sharing of information in a collaborative way. I explore how emerging patterns of collaboration within the context of volunteering are bound to the affordances of the tools. My initial idea was that collaboration was seemingly emerging from the affordances of digital tools embedded in volunteering processes.

The concept of affordances has been explained by Billet (2001) by providing an example of a door handle. The door arm possesses an affordance for being twisted in the direction of the clock and pulled or pushed by a right handed person so that the door can be opened. Due to these attributes of its design, the handle provides support for the individual for perceiving what kind of actions are to be made with the object (Billet, 2001). Similarly, Norman (1988, p.9) uses the example of a chair in the sense of its support for sitting and defines affordances as the perceived traits or attributes of a particular thing which specify the user actions to be undertaken through the means that thing.

Several affordances can be mentioned for the technologies used for learning. For example, typing and editing posts are related to blogging, yet these are not affordances on their own. They provide support for the affordances of idea sharing and interaction. Within the context of interaction Hartson (2003) made a distinction between affordances based on the following four types: cognitive (thinking), physical, functional, and sensory. One of the underlying affordances of learning technologies, specifically those that fall into the category of online collaboration tools, seems to cut across all of Hartson's categories—the social affordance of these digital tools. Consequently, the term 'affordances' is used here not to refer to one of the four categories mentioned above, yet to refer to the social affordances of the tools used throughout the CSC Program.

Overall, I found out that whenever the affordances were richest, the related collaboration results were higher. The online CSC environment was highly invitational for collaboration that was also confirmed as such by the volunteers. The openness of these applications for constructive interactions also belonged to the affordances. Yet, it should be kept in mind that the affordances of these tools on their own cannot automatically result in a high level of collaboration when individuals decide not to participate in these online experiences.

### 4.4.3 Properties of Technologies that Influence Decisions

One of the features that was most frequently mentioned by CSC volunteers in terms of the impact of a particular tool on decisions about both the use and its underlying nature, was whether or not it was a technology that volunteers perceived to be support for socializing with others. Due to the multifaceted nature of the interactions in the CSC Program, participants equipped with various levels of expertise and areas of interests nourish the volunteering environment with ideas and knowledge that are befitted by volunteers based on their needs. By using the available tools, they have the opportunity to support each other, justify their views and opinions, and offer suggestions and explanations. Also, volunteers acquire a greater level of expertise as their conversation via these tools got enhanced.

Participants have an awareness of the gains or pleasures that they might be provided due to the usage of these technologies, yet they have to decide on their own whether the perceived beneficial properties (learning or social) outweigh the costs:

> "One of the downsides of using online collaboration tools for professional knowledge-building and sharing is that they cannot be used for arriving at a precise decision. Very often the discussions tend to be elaborating and not focused at arriving at a conclusion. Someone needs to be a moderator in the discussions and it takes time to find who could be the best person to do this when compared to a face-to-face group interaction." CSC Participant [12]

CSC participants view these technologies as having useful properties to support their process of collaboration, but make refined and complicated decisions about their simple usage, based on some of their properties such as how engaged they feel with using a particular tool (Collis and Moonen, 2001). During this decision-making process, sometimes participants perceive they are involved in a subtle balancing act; sometimes participants have the feeling as if their choice is rather limited:

> "in terms of socialising, you choose what you want. While for collaboration, you have to take what is available." CSC Participant [10]

Moreover, some participants mentioned that the properties of these online collaboration tools for a social activity such as CSC should also be utilized for other projects.

> "I would love to see online collaboration become more of a way of working, versus something you do for social activity such as the CSC or when there's a 'collaboration' project. I try to be a stimulus for change by sharing material; however, I tend not to go back if others don't engage. What's the secret to bringing an online community to life at all?" CSC Participant [1]

Some participants also mentioned that the ability to save time is another important property of these tools.

> "Especially time- time in the long run is such a central point. [...] I think that it is almost always worth putting in that additional effort to choose something up that will save you in the longer term. But it's the delicate act - It's finding the time to leaves and do that." CSC Participant [13]

One important point to conclude from these statements is that the personal benefits that volunteers perceive as gains from making use of the properties of these tools are more important than the intrinsic features of any particular medium.

In summary, volunteers found that various properties of these tools offered them many possibilities in terms of providing an opportunity to get involved in various different communication processes based on their individual needs and preferences. A particular property offered by a specific tools is not simply an 'add-on', it is central to how the volunteers organized their volunteering experiences by offering alternative routes to participation responsive and immediate modes of communication and interaction, which makes their work and volunteering life to much more flexible.

There are also some volunteers who do not choose to use the collaboration technologies. In the next sections, I will present the personal accounts of these participants.

## 4.4.4 Decision for not Using Digital Tools

Only one participant shared his statements about his decision for not using any digital collaboration tools (excluding e-mail; Participant 9) although most of them also mentioned that they would prefer face-to-face communication to most of the tools available. For Participant 9, the decision was made due to the fact that he did not feel any particular need to use them:

> "I'm not used to use online collaboration tools in my work." CSC Participant [9]

This decision was influenced by the fact that most of the time, he worked with his team members closely and even e-mail and phone calls were enough for him to be in contact with them. The fact that other team members gave up using some of the tools also contribute to his decision for not making use of these tools.

> "I think that this travel can be described as a tour for listening. My CSC teammates and I interviewed several academicians, officials from the government and those working for non-profits as well as participants in the livestock economy. During my conversations with different people such as farmers or traders it was surprising for me to find out that no one had an idea about Market Information System of Livestock (LMIS) which was operating within the country for three years. The system aimed at developing the transparency of the markets. This made me realize an important lesson: Even though you may create something useful, people may still not know how to use it. I work for marketing, so, it was satisfying to discover how crucial that project can be." CSC Participant [13]

This IBM volunteer was involved in a marketing assignment and as part of their group project a Market Information System of Livestock needed to be developed to contribute to the livelihoods of farmers. He realized that only by building relationships with farmers in his local project community

70  *Results*

he would be able to convey the benefits of the new system. So, apart from creating shared repertoire by mutual engagement, bonding and culture are also important for creating a shared identity for volunteers. Bridging (weak ties) and bonding ties (strong ties) are required to provide support and create value in the network (Putnam, 2000). This is also in alignment with the following remarks:

> "Within this web, there are relationships referred to as bonding which isled by feelings that entails a group of individuals - relationships which reinforce each other, rather than a mere chain of one-on-one relationships.... Moreover, some type of commitment should exist with regard to a group of common values, a shared history and conceptualizations—in other words, a culture." (Etzioni, 1975, p. 241)

These statements are in contrast with Orlikowski (1992)'s hypobook which claims that these technologies will be underutilized if people don't appreciate the collaborative nature of online tools despite the existence of shared values. Although the CSC employees have an awareness about the socio-technical capital provided by these tools in the form of shared history and values, due to the privacy and lack of a clear social media policy they may not always use these tools to their highest potential.

### 4.4.5 Preference for Using Certain Digital Tools (Driven by Desire)

Some volunteers mentioned how they were searching or hoping for a certain type of digital tool that would fulfill their requirements for collaboration more effectively. Their decisions or preferences were typically grounded on both a good working knowledge with regard to the potential benefits of these tools and extensive experience of the challenges with regard to the use of these tools more to satisfy their desires.

Participant 16 also mentioned that encouraging everyone to use the tools and letting them know the importance of using the tool is important. In addition to the sharing of these stories about volunteers' engagement, Participant 16 also mentioned that volunteers' desire to participate in online collaboration platforms is also due to their ability to infuse flows of intelligence for extraordinary collaboration by making use of these tools. According to Participant 16, some underlying desires of the participants for the use of these tools are:

- To find meaningful insights
- To decide about the role individuals need to play

As Participant 16 mentioned, one of their desires is the clarification of mutual roles and responsibilities which is very important with regard to the use of web-based tools (Westera, 2004). This interdependence between individual roles and responsibilities and volunteering communities supported by technology provide a range of option to employees to fulfill their personal needs and achieve their goals. This recapitulates the underlying principles of the Web 2.0 era—that at core of the Web lies establishing connections between ideas, minds and communities, while facilitating individualization and collaboration resulting in joint knowledge creation.

Another desire mentioned often was the participants' desire to 'stay current' as Participant 17 mentioned:

> "We should specifically put effort not to miss anything a team member might have delivered by checking recordings of conference calls, meeting minutes, and other ways to stay up-to-date in order not to lag behind." Participant [17]

As these statements emphasize, the CSC Program exists within a socio-cultural system in which participants utilize different tools and various forms of interaction to constitute a desired collective activity, supported by the affordances of the digital tools. The collaborative tools used symbolize 'affinity spaces' which is defined by Gee (2004) as spaces for the acquisition of social and communicative skills, as well as engagement in the participatory culture of Web 2.0.

## 4.5 Evaluation of Support

By undertaking this study, I as a researcher also delved into the digital tools and social processes necessary for completing the kind of distributed volunteering work under scrutinization here. Almost all of the CSCL literature suggests that the right mix of digital tools is crucial for aiding the development of online communities (Boyd, 2007; Kiesler et al., 1984). More complicated and media-rich online spaces including audio, video, instant messaging, and alike may contribute to the minimization of any differences between digital and face-to-face environments (Kiesler et al., 1984). Yet, due to the transitory nature of online communications, individuals may have inflated expectations for prompt feedback and got frustrated in case that does not happen (Kiesler et al., 1984). In this section, I will expand on the following identified themes or issues:

- Reasons for not receiving any technology support;
- Evaluation of informal sources of support.

### 4.5.1 Reasons for not Receiving Any Technology Support

Although the volunteers did not face many technical issues when it comes to the use of online tools they still mentioned that it was easier to access the colleagues in the case of this type of issues or online forums were also another alternative for receiving support.

#### 4.5.1.1 Hit and miss

For some volunteers, a preference for an experimental approach such as hit and miss or trial and error showed the level of confidence with regard to their technological skills. The tools extend the range of the volunteers allowing them to complete a task that would be otherwise difficult; and they are utilized to selectively aid the volunteers in case of any need to do so.

> "When I make a trial with a new application or tool, I start thinking whether right 'now' I can make use of it, and try to decide to attend a related training on it or not. I prefer to spend an hour playing around with it and exploring it in detail, rather than spending more time on related trainings." CSC Participant [12]

As individuals decide whether to accept or reject a particular tool; they tried various different tools before having to commit to the use of a particular tool and evaluate then in terms of the relative advantage over other available tools, observability of its benefits and complexity or ease to learn. For the CSC participants, apart from the perceived benefits of online collaboration tools, the trial ability also made these tools embedded within the practice of volunteering.

Based on the findings it is apparent that volunteers' participation in digital collaboration should not be assumed as being certain. Even when support is provided which makes these tools more invitational, individuals may prefer not to put effort for any online participation or appropriate the existing knowledge based on their own needs. Meaning and value are important for what is afforded for them to participate in online conversations and learn. So, there are different kinds of invitational qualities required such as the ability to make reluctant participants get involved in conversation and support them for finding a meaning through their participation in ways that enable the transformation of existing values and practices.

### 4.5.2 Evaluation of Informal Sources of Support

As the CSC environment was technology enriched, the notion of support also evolved; the concept of support goes beyond the interactions among individuals as artifacts, environments and resources all of them are utilized for the aim of providing support. Based on a repository of approaches and methods, the volunteers adjust the level of the support depending on their ever-changing set of knowledge and skills on an ongoing basis. Thus, the quantity and kinds of strategies differ not only among different volunteers with various levels of expertise, but also for the same volunteer over a time period.

Most of the participants reported that general training about CSC collaboration tools was good, two of the participants, however, could not be so sure about the quality of support they received and had mixed feelings. For instance, Participant 9 mentioned that getting valuable insights and information from other fellow members using the tools, and discussion of improvement ideas for the program/project were other ways of getting support. So, apart from the existing support available via online tutorials or training, peer-support (Stephenson, 2003) was also mentioned as being significant for the participants. Participant Twelve also emphasized that the help that one needs has to be immediate. Yet, the degree to which the social network augments or takes over from the available mechanisms with regard to providing support is not obvious. Volunteers still gain benefit from the conventional support mechanisms such as project materials with regard to organizational workshops and training: the extent to which some of these are taken over by volunteers from the guidance of fellow volunteers for support might need further exploration (Figure 4.4). One of the most compelling examples with regard to the timely use of the peer network can be exemplified by one of the volunteer voices:

> "The biggest down side with Lotus Communities - it's easy to assume that someone has seen or read an update that's posted into the community web site, when in fact that might now. As an example - I had a problem with my notebook connecting to the internal network for several days. During this time a few team mates

## 74  Results

**Figure 4.4**  Overview of Lotus Notes collaboration space.

thought I had seen a document that was posted when I had not. There were confusing conversations (not helped by language barriers) that were only clarified once it was clear that I had not seen the actual document. Eventually, we sorted it out." CSC Participant [8]

While we transfer the notion of volunteering into technologically enriched environments, the notion of customized support becomes more difficult. Support has therefore been transformed into various formats along with several affordances. Online environments consisting of many individuals do not enable the sensitive, customized and delicate exchange that happens in a one-on-one situation (Rogoff, 1990). Therefore, support is now being conveyed in the form of an application or tool with which the user interacts. Instead of one experienced person giving support, volunteers now have a network of distributed support experts in which volunteers scaffold each others' resources whereas the learning environment itself is reconstituted to enhance the level of both motivation and support provided.

## 4.6 Understanding Perceptions and Beliefs of the CSC Participants

One of the main findings is that participants' usage of collaboration tools is multidimensional, sophisticated, and customized based on individual needs. Volunteers are provided with information from various knowledge sources and consequently, both the ways in which volunteers use information and communicate appear to be changing. All these tools provide the volunteers with the option for controlling their own learning and directing the voice of their related CSC group through means of a raft of tools.

### 4.6.1 Perceptions of Technologies and Their Use

The participants are generally sophisticated users and see technology as integral to their collaboration. They are also aware of the opportunities to build a robust and solid amalgamation of knowledge from nascent and discontinuous information.

> "In my opinion I don't try to customize the tools based on related volunteering activities, rather I use them as complimentary i.e. searching the wiki to find specific information and then asking my colleagues for it using e-mail [...]. In this age, we are using new technologies all the time, and starting to use new ones whenever new technologies emerge, it's not a case of "customization"; it's just the way we work, using multiple methods, some 'traditional', some digital." CSC Participant [13]

This participant also found the concept of distinguishing between 'collaboration' and 'online collaboration' irrelevant.

The perceived benefits of online collaboration tools can also engender 'epistemic fluency' (Goodyear and Zenios, 2007, p. 358). Epistemic fluency provides one with the opportunity to realize, grasp and appreciate the complicatedness of a community that one has not met before, regardless of the fact whether this community is religious, professional, or scientific (Goodyear and Zenios, 2007, p.358). From the perspective of the professional CSC community, knowledge-sharing goes beyond the embodiment of existing ideas, norms, and practices; it involves qualitatively another level (Engestrom, 2001).

The main perception about the Web 2.0 tools were their function as being intermediaries for producing learning artifacts. Volunteers made use of these tools as a mixture of different functionalities offered by these tools in alignment with their own individual needs. Despite the fact that the use of these tools was almost pervasive, the ways in which individual volunteers utilized the tools varied to a great extent. For example, the volunteers used the wiki not only for delivering project reports, but also on a wider basis – for example when working on individual and group tasks, adjusting a report and reading project-related material. Moreover, based on the findings I can say confidently that the wiki is being utilized at a very sophisticated level. Due to the opportunity of writing down ideas and then making revisions or to bringing ideas together a major change in the practice of volunteering takes place. All these evident patterns for collaboration may also indicate the co-creation of knowledge arising from these interactions between the volunteers and the tools dispersed across several digital spaces, e.g., Salomon's Notion of distributed cognition. This also evidently points to a transformation from old concepts underlying the importance of volunteers concentrating on single sources of information into the multi-faceted nature of their knowledge work.

### 4.6.2 Beliefs about the Value of Online Collaboration Tools

In order to make a more in-depth inquiry about the positive feelings underlying the use of digital tools I will delve into the following themes:

- Technology offers an easier life for the volunteers (this statement is related to the themes of beliefs about ability to collaborate without technology and beliefs underpinning the value of technology for contributing to the process of collaboration);
- Positive statements with regard to online collaboration tools (for the purpose of discovering the volunteers' views about the value of technology in terms of its support for various kinds of collaboration; also see Section 4.6).

#### 4.6.2.1 Technology provides ground for easier collaboration

Technology provides ground for easier collaboration according to the statements of some CSC participants as it offers an opportunity for easier access to more relevant information at the first place. For some participants, there has been so much information that their progress would not be possible without the means of technology. For others, technology is not so great as it does not always work. Also, the value it provides should be clear.

## 4.6 Understanding Perceptions and Beliefs of the CSC Participants

"I have nothing against the use of technology support my learning, as long as it functions in a useful way. With the use of Sametime or IM, there is a potential for too many interruptions. The IBM blogs that we used can definitely be improved, make it more user-friendly. The important question to ask should be which kind of value to create by contributing more to this area. If there were an obvious answer to that question more could be invested into this space." CSC Participant [10]

"One of the first things I learnt about is how important it is to get prepared. As my project was about cloud computing technologies I had a gamut of terminologies to grasp. Every day, I woke up earlier than others to get prepared for the project meetings. There were even jokes among us such as "I even did not work for SAT so hard." Yet, the preparation turned out to be worth it as I was feeling confident sitting in front of the director.

Also, another important thing I learnt about is that Google Translate can never replace a good interpreter as there could be many cases in which the relevant and non-sensitive word needs to be selected. Of course, with regard to learning Chinese I have a long way to go, yet apart from the language I also learnt about interpersonal relationships, negotiation and coordination. CSC Participant [9]

These statements above imply a connection to Karahanna et al. (1999) assertion that the utilization of online collaboration tools drew on a wider spectrum characteristics than continuing users, who focused on their perceived traits that make life easier for them. It is also important that groups are working toward providing more value for others (Kasworm, 1997).

### 4.6.2.2 Positive thoughts about how online collaboration tools are used

As stated earlier, when asked about the positive thoughts related to the usage of online collaboration tools participants stated that they see most value in terms of the social affordances of these technologies (See Section 4.6).

One participant specifically mentioned the opportunity provided by the online collaboration tools to go through a team evolution process and recognize common patterns in communication styles of group members.

"The use of the collaboration tools is unique in that it will provide for both IBM CSC volunteers and local communities to learn from each other and tell their stories about their building a Smarter Tianjin." CSC Participant [5]

"IBM takes skill-based volunteering very seriously. As IBM employees around the world help local communities during the CSC Programme they also find the opportunity to use their various skills. They all share the same objective of using their expertise for being helpful for others. Their volunteering roles range from being a fire fighter to teacher or technician." CSC Participant [5]

"I really enjoy crafting a story. It helps me put a pile of factors into a cohesive picture and articulate my view. In fact, this is something that energizes me at work as well, so the subject matters less than the process. (Note that in our client work, I found I excel at creating the logical and theoretical stories.) [...] We each impact people without knowing it. As we shared our personal findings in the end of the program some people shared unexpected and significant changes in their outlook for their future and even beliefs. For every shared change I know there were many untold.

Sharing our vulnerabilities can help others. There can be risks with this for sure but when in the right context, opening up can be inspirational and encouraging to others while feeling darn good.

Lasting relationships are great; so are meaningful moments. I had some great moments with people I may never see or speak to again. When I consider it, I realize the relationships I maintain are based on the meaningful moments within them.

These aren't all new lessons, but they are renewed with different analogies and experiences. Relearning a lesson can be as important as learning it the first time. CSC Participant [5]

This participant further elaborated that technology plays an important role in forming teams. There was also the mention of rotating leadership among team members through the use of these tools. Participant 5 mentioned that capturing the CSC best practices via an online platform is really useful. This participant also mentioned that they rotated the discussion forum leadership around the team, and had the volunteer scheduled to lead the next topic be the scribe for the prior discussion on the wiki. As a result, most of the volunteers would be sure to be dependent on contributing more, since it would affect the agenda of their discussion.

## 4.6 Understanding Perceptions and Beliefs of the CSC Participants

"Just to level-set everyone. I'm here with 12 pretty amazing IBMers, 4 from US, 2 from India, 1 each from Germany, Hong Kong, Italy, Australia, Canada, South Africa. We are divided into 3 teams of 4, working on projects with different Government agencies. A Canadian NGO (DOT, http://www.dotrust.org/) is managing our relationships and our projects. [...] So that point is something I have wanted to blog about since Sunday when we had our first meeting with our DOT host. Kenya has a new constitution, a new president and a plan for where they want to be in 2030 (Vision 2030). Over and over again, the Kenyans we meet demonstrate an incredible sense of optimism and energy for the future. Being from the US, it seems to me that this is how Thomas Jefferson and John Adams and their peers must have felt in the early days of our nation." CSC Participant [10]

"In order to add value in service to local communities, there is an ongoing collaboration between IBM employees and non-profits via means of our On Demand Community, Corporate Service Corps or Smarter Cities Challenge teams" CSC Participant [5]

As these statements demonstrate, getting an information group going requires a critical mass of contributors and contributions, leaders who start by contributing more than they get in return (altruistic contribution), trust in how contributions will be used, reward and recognition for contributions. While IBMers take on the role of helping others, they also act as e-facilitators as the work with the ideas presented to synthesize and promote further debate and clarification. Haythornthwaite (2008) refers to e-facilitators also as "braiders" based on the term "braided learning." According to Haythornthwaite (2008), braided learning is a theory based on observations of learning within communities of professionals. There are three phases in the process of online interactions that provide evidence that this particular kind of professional learning is taking place. The first phase is when the community gets involved in submitting a braided text online which can support different opinions and changes of mind. Some volunteers take the role of e-facilitators or braiders who facilitate the shaping of the argument, preparation of draft summaries and changes based on the suggestions provided in the discussions. In the second phase, these facilitators or braiders display higher levels of development by building braided artifacts that again interprets the virtual debate in different ways for different target audiences. The third phase of learning appears when braiders establish groups to make an in-depth inquiry about a particular subject. A gradual evolution into becoming a pro-active professional occurs by making use of the knowledge in a collaborative manner to create new procedures and theories that will ultimately influence their expertise in the near future.

Based on these stages of development, the group processes can be summarized as follows (Haythornthwaite, 2008):

- Offer ideas and alternative plans
- Choose a plan of preference among the alternatives
- Make negotiations amidst conflicting views and interests
- Implement the work by applying some standard to it given the competitive environment

While the main benefit can be cited as active construction of knowledge, enhanced problem articulation, peer-to-peer information sharing, trying ideas out on others a need for both learning and for a safe feeling in the expression of the unformed ideas occurs. The ultimate objective of collaboration is to build a community of inquiry in which users participate to construct meaningful and worthwhile knowledge in collaboration.

> "When I arrived in the country due to my project a few months ago I came across an article on the plane about Gandhi which reminded me of one of his famous quotes about living as if one were to die tomorrow and learning as if one were to live forever. Keeping that in mind, I started my CSC assignment![...] Conveying my know-how to a highly visible national project that supports African communities was surely very satisfying, yet the professional know-how I gained was also very impressive I think. Collaborating within a developing country and gaining an understanding about its areas for development is a mall step on my learning journey. [...] During the IBM CSC, an opportunity to make a difference in the world is provided to any participant. Such an amazing adventure for sure!"
> CSC Participant [13]

While I mentioned the term "community of inquiry" here it is essential to mention the following differences of meaning among different types of communities (Kazmer, 2007):

- Community of Practice (CoP) denotes a shared area of work and the main emphasis on practices.

- Community of Inquiry refers to a shared area of exploration while interaction occurs by mutual construction of meaningful knowledge.
- Knowledge Community refers to shared knowledge based on particular expertise where specialists are distinguished from newcomers and members from non-members
- Discourse Community refers to shared genres, communications with specific goal and shape in which community members make use of traditional means to achieve social goals, e.g., rituals for greeting, forms of online conversation. The level of knowledge with regard to a particular domain determines membership.

These statements suggest that establishing a commonality is the major prerequisite for a successful collaboration in alignment with the suggestions provided by Olson et al. (2002). According to these scholars, the main component of collaboration is rooted in the notion of coming together for work with the purpose of obtaining a particular objective. A sufficient level of commonality must exist for meaningful and collaborative interactions to take place (Olson et al., 2002, p. 49). The volunteers tried to build this commonality by conveying the best practices through the means of these tools. This differs from the common understandings of commonality which take the individual unit of analysis as a starting point and then look for an explanation of a shared reality. Commonality seems to refer to a shared understanding among individuals who are equipped with knowledge based on their professional and educational level which enables further interactions (Stahl, 2001). But, throughout the CSC project, the meanings are constituted within an activity system that provides ground for individual interactions. So, the commonality is not available for them from the beginning onwards, yet it has to be developed through some kind of agreement on a mental level. The commonality was interactively achieved; so its construction should be attributed to the group interaction via use of online collaboration tools rather than to any kind of spontaneous agreement among the individuals to the ideas acquired previously within their individual minds. The volunteers had to focus on the tools as a solid artifact and interpret its meaning. The ability to make these digital artifacts an important aspect of their volunteering activity was a goal that they had to reach in a collective manner.

### 4.6.3 Mixed Feelings about the Value of Online Collaboration Tools

Mixed feelings arise due to the sophisticated awareness of the CSC participants. These feelings entail both positive and negative feelings. Some participants expressed beliefs that they wouldn't be able to collaborate without technology, yet expressed some issues of concern such as privacy. Still some of the volunteers stated that they would collaborate without the use of digital tools, but they would prefer to still use them.

For some participants, time was the most important aspect and more tools to use meant more time to spend on them. For others, security was more important. Participant 2 mentioned that security is what mostly concerns any organization to allow the use of online tools. Yet, still, he thinks that for a global team these tools are the best and always beneficial.

Some volunteers mentioned that the absence of clear social media policy guidelines causes also an obstacle for the effective utilization of these tools.

> "[…] Privacy is my main concern when it comes to collaborating via online tools. Less obvious is IBM's clearly defined social media policy which provides guidelines and allows us to use non-IBM tools to write about CSC. I think it touched on that already but it's something that needs to be considered and that's a big part of IBMs social networking guidelines." CSC Participant [8]

All these different statements demonstrate that the mixed beliefs of the participants about the usage of digital tools is also related to the shift in the organizational culture (Rosenberg, 2001); therefore there are different opinions on whether these tools are providing value or not.

### 4.6.4 Defining Appropriation and Influential Factors for Intersubjectivity and the Practice

The term 'mutual appropriation' has first been used by Rogoff (1990) to refer to the multi-faceted nature of the interactions involving the communities of learners approach. Central to this concept are interaction, collaboration and negotiation are at the core of such a context. The same concept can also be used within the CSC context as the "appropriation" of volunteers' thoughts in such a context has many dimensions and they appropriate ideas and skills by making use of various tools. At the core of successful collaboration lies a shared understanding of the goal of the task. Although some elements of the task may be beyond what a volunteer could complete on his own, a shared conceptualization of the task or in Rogoff's (1990) terms "intersubjectivity" has an important role to play. Intersubjectivity is obtained when there is a shared ownership of the activity and a common conceptualization of the

### 4.6 Understanding Perceptions and Beliefs of the CSC Participants

objective as a result of the collaborative redefinition of that activity. So, perspectives are negotiated on an ongoing basis.

A data analysis based on the three sources puts forth a number of unexpected results which provides a unique insight into volunteers' existing practice with regard to the use of these tools and their experiences to attain intersubjectivity. Based on the emerging data, I discuss these findings under the following main headings:

- Appropriation of technologies by volunteers;
- Changing practice and volunteer competences.

#### 4.6.4.1 Appropriation with regard to digital tools and activity types

The results of this study made it evident that participants were putting the tools into practice to contribute to all aspects of their volunteering process; ranging from communication and collaboration to presenting and reflecting upon their ideas.

As the notion of the community was getting lost in even the less developed world such as Ethiopia in which IBM employees were contributing to the project through use of online communities, we should perhaps bemoan the approaching end of the ideal community in the form of a village, considered lost in urbanization. On the other hand, a more liberated definition of community occurred derived from the idea of social networks with an emphasis on social relationships rather than geographic location. These virtual communities are place independent, liberated from geography and dependent on technology (Haythornthwaite, 2008).In CSC Program, online interaction support offline geographically dependent communities whereas online interaction and engagement go hand-in-hand with overall civic engagement. Also, by using information and communication technologies to improve communities community informatics becomes more important.

Furthermore, the study showed that their use of digital tools for collaboration is also related to their usage for the purpose of social activities. In this section, based on the findings discussed in previous sections I provide descriptive examples of how the IBM employees are using these tools throughout the CSC Program (Table 4.1).

1. Communicative Tasks:
   By communication, participants refer to other aspects of a conversation as well such as providing feedback, listening and other activities that underlie having a more open relationship with others. Rather than

underestimating the importance of any of these components, the focus should be on which channels might be the most effective for conveying a particular message. After all, both leading and learning requires one to listen as Participant 17 mentioned.

Examples of the dialogic nature of tools are prevalent throughout the data. Some participants also admitted that think the diversification of the communication tools that are also being championed by the management sets the CSC Program apart from other employee volunteering programs.

Participant 17 mentioned that by being able to listen to others one gains access to information that can be acted upon and feedback that gives form to one's interactions. He also asserted that through listening, one can get informed about what others might be concerned about or find value so that one can increase his participation in a more unique way. Similarly, Participant 16 stated that the important question to ask should be which available tool to use rather than to use it or not at all. It is also a question about desiring to share the stories online:

> "The opportunity for organisations to publish their own stories of community engagement has never been so big. Due to the available social networking tools like Facebook, Twitter and LinkedIn companies, employees and even customers can easily share meaningful stories of community engagement through various, dynamic spaces. [...] Without these tools of these stories could never heard of. Instead, they would be forgotten amidst the chaos of everyday life and work." CSC Participant [16]

### 4.6.4.2 Types of learning

As mentioned earlier, IBM's approach to the CSC employee volunteering program is regarded as an informal type of learning activity. Embedded within the program were various online collaboration and learning activities. Therefore, a review of key themes related to online collaboration is important in order to understand these activities better. Dyke et al. (2007) offers a review of digital learning theories and states that there is need for a metaview of the main topics cutting through these different positions. These authors state that non-formal learning could be enhanced by the focus on the following three main elements of collaboration:

- through reflection;
- through conversation and interaction (Dyke et al., 2007).

## 4.6 Understanding Perceptions and Beliefs of the CSC Participants

Within this context, what is apparent from the data is that most of the volunteers take these aspect into account in terms of combining the tools, intermingling and re-appropriating materials, controlling their meaning and developing a shared understanding with others based on self-reflection and evaluation.

1. Reflection
   
   The data shows that technologies provide volunteers with numerous opportunities for reflection – tools to mix and match information, the ability to compare news from different online f sources, and tools to support their collaboration. Various examples have been mentioned in the data with regard to reflection being an integral part of the collaboration process implicit in their volunteering practice such as the cathartic nature of their blogging practice.
   
   The data further reveals that the support and openness for online collaboration also have an impact on the reflective process embodied in their collaborative learning tasks. In some ways, these findings are based on common sense. The kinds of opportunities made available for volunteers to reflect on their tasks through the use of online tools will be crucial for the quality of emerging collaboration. In a similar vein, how individuals get involved in collaborative practices will shape how and what they learn. Nevertheless, these factors may not be taken into account in case the relationships between engaging in reflection and learning through collaboration are not fully understood.

2. Conversation and Interaction
   
   Based on the main principles of social constructivist learning, effective learning is conversational in nature, and it requires a social dimension in the form of communication, dialogue and shared activity. Dialogic activities can be recognized around all aspects of the reported volunteer activities. All statements cited so far can actually be mentioned here again as without conversation and interaction, the practice of volunteering cannot occur. For the sake of not repeating them, I briefly include some other examples as provided by the CSC volunteers:

   "We should select and reinforce a particular model for collaboration. An effective collaboration can't occur if team members are not provided with a role model in each group for getting out of their comfort zones. If, on the other hand, there are one-on-one conversations with this role model, and this role model is privately answering questions from individuals that should be shared with the all other team members, your team will be inspired by this model as well." CSC Participant [17]

The tools helped volunteers to get exposed to important ideas (Reiser, 2002) and supported the externalization of processes that would otherwise remain tacit. Their collaboration process has also been transformed via a careful arrangement of the environment – both online environment and the physical volunteering environment.

There are two important conceptual implications based on the specification of these kinds of collaborative learning. Firstly, with regard to deliberation within constructivist theory a conceptualization in terms of the affinity between participants and social practice is crucial. Here, it becomes evident that individual actor exists both independently and interdependently in social practices which is also in alignment with Engestrom and Middleton's (1996) proposal. So, as Hutchins (1991) mentioned, the individual does not simply constitute an aspect of social practice and this actor realizes himself in various kinds of ways. Evidence of both interdependent use (utilizing the tools in order to collaborate), independent use (e.g., independent use of tools for other purposes than collaboration) are evident throughout the study. Such practices do not often conform with the norms and practices of the conventional volunteering practice. Individual traits such as personal values act as a mediator for the methods of collaboration and learning within the context of social practice, such as virtual environments. Linkages among these individual values and digital practice give form to the individual engagement within the collaborative practice of volunteering. The types of collaboration as determined in the CSC Program marks the start of the coming to a realization of the range of different relations between the individual and social practice which eventually shape individual paths to learning with these tools.

Second, due the affordances of these tools to offer access to crucial knowledge related to certain subject areas, it is important that these tools be highly invitational. According to the findings in order for online collaboration to progress effectively, the way the volunteers are enabled to participate in online conversations and be supported in this progress will form rich learning outcomes.

### 4.6.4.3 The shifting volunteering practice
The findings mentioned above suggest a shift in the way in which employees are volunteering and point to a complex and rich inter-relationship between

## 4.6 Understanding Perceptions and Beliefs of the CSC Participants 87

the tools and the users. The extent to which IBM employees are possessed with technical competencies obviously has a crucial impact on their adoption of these digital tools.

The following factors as determined previously by Cogburn (2002) with regard to collaboration also are prevalent in the process CSC volunteers' collaboration and they point to the changing practice of employee volunteering:

- Collaboration readiness: Book suggests that a high level of collaboration readiness is embedded within these volunteering projects. These collaborations have developed their own rules and procedures according to the volunteering tasks at hand. Collaboration tools potentially allow more volunteers to directly participate in collaboration with other volunteering groups. Some degree of commonality is required so that successful interactions can take place. Participants who share more commonalities will find the communication process easier (Clark, 1996). This becomes even more crucial when a high degree of trust or negotiation is required for the interactions. According to related research in this field, in order to eliminate lack of trust development, an extra effort should be put for establishing commonality before using the digital tools (Rocco, 1998; Zheng et al., 2002).
- Collaboration infrastructure readiness: An obvious challenge in these volunteering tasks is working with the existing collaboration infrastructure in less developed or developing countries. Despite the availability of the necessary bandwidth, communication networks seem to be unreliable and might be overloaded at different times during the day. Frustration may arise due to these network failures when reaching collaborators is essential for the work to progress. There is a least likely opportunity for the usage of these digital tools in case of the unavailability of the network.
- Collaboration technology readiness: In addition to the provision of digital tools that fulfil the current level of collaboration technology readiness, an adequate level of readiness among the volunteering groups should also be established. There is the requirement for the participants to establish a common understanding and facility with digital tools before these can be utilized for getting successfully engaged in distributed volunteering.

All these factors lead to the emergence of a new volunteering practice that I call as 'distributed' or 'technology-enhanced' volunteering. It is the amalgamation of the social affordances of digital tools, with new informal learning goals and priorities that provide an opportunity for metamorphical shifts in employee volunteering practices. Based on these findings I am optimistic that the new tools will result in employee volunteering programs such as the CSC Program that are more personal, participatory and collaborative.

# 5
## Some Initial Reflections

There are many potentially intriguing findings in this study (Tables 5.1 and 5.2), however, one should also bear in mind that the measures are based on self-reports and the perceived benefits of utilizing these tools and their existence might not be in alignment.

To begin with, collaboration tools are more than a mere piece of application. During the CSC Program, they are situated in the practice of volunteering and include people working toward the achievement of a shared goal (Wenger, 1999). The CSC Program includes an informal curriculum to be completed as well as the volunteering practice in which project work is based on various learning activities. These dimensions enhance the meaning of the use of collaboration tools and shape the volunteering practice that take place.

As any other group, the volunteering project group is made up of individuals with their own personal lives, thoughts, beliefs, and aims which is a prerequisite for project-based learning according to the constructivist perspective. Individuals carry their own acquaintances into the research and a shared agreement is obtained by means of confrontations and negotiations of perspectives and beliefs (Dirckinck-Holmfeld, 2002). Collaboration among the volunteers is an important requirement for project based learning so that the mutual engagement and interdependency can be facilitated and maintained in order to undertake a successful project. Going beyond the coordination of tasks and meetings the collaboration involves the whole mutual process of structuring and establishing a common understanding.

The case study revealed two dimensions of collaboration that are evident in the CSC volunteering projects: Collaboration with regard to the CSC Program activities and collaboration with regard to knowledge construction. Collaboration with regard to CSC Program activities is more an administrative task while collaboration with regard to knowledge construction has to do with the mutual negotiation process of obtaining a common understanding throughout the project.

**Table 5.1** Mapping between the first research questions and the findings

| Research Question | Findings | Cross-Reference | Sub-Themes |
|---|---|---|---|
| How are collaborative learning tools used for the volunteering practice of knowledge workers? | While connection is about enabling a space for activities co-presence is about requiring that everybody participate in these activities. | Section 4.1 Figure 5.4 | Usage and Approaches Strategies/Choices Feelings |
| | Informal and formal distributed cognition are apparent throughout the CSC project. Through participation in these forms of discussion and interaction, volunteers are provided with the ability to construct their own informal learning trajectories as well as shaping pro-actively those of others. This observed distributed cognition among CSC volunteers as supported by online collaboration tools directly leads to the temporary construction of one or more group minds. | Section 4.2 Section 4.3 Figure 5.1 | Usage and Approaches Strategies/choices |
| | Intersubjectivity is obtained when there is a shared ownership of the activity and a common conceptualization about the objective as a result of the collaborative redefinition of that activity. So, perspectives are negotiated on an ongoing basis. | Section 4.6.4 Figure 5.4 | Usage and approaches |
| | The perceived benefits of online collaboration tools can also engender epistemic fluency (Goodyear and Zenios, 2007) which allows volunteers not to underestimate the complexity of existing ideas, norms, and practices. | Section 4.6.1 Figure 5.4 | Usage and approaches Feelings |

| | | |
|---|---|---|
| Evidence of both interdependent use (using the tools for the purpose of collaboration), independent use (e.g.: independent use of tools for other purposes than collaboration) are evident throughout the study. Individual traits such as personal values act as a mediator for the methods of collaboration and learning in social contexts, such as digital environments. Interrelations among the values of the participants and the digital actions give form to the individual engagement within the collaborative practice of volunteering. Meaning and value are important for what is afforded for them to participate in online conversations and learn. So, there are different kinds of invitational qualities required for finding a meaning through voluntary participation in ways that enable the transformation of existing values and practices. | Section 4.6.4.2 Figure 5.1 | Usage and Approaches choices |
| Given the situatedness within the context of relations and distributed volunteering networks, the volunteering experience requires a certain amount of commonality in order for collaboration to take place. | Section 4.6.4.3 Figure 5.1 | Usage |

Table 5.2 Mapping between the second research question and the findings

| Research Question | Findings | Cross-Reference | Sub-Themes |
|---|---|---|---|
| What are their beliefs about the benefits and challenges in using these tools for such a practice? | The main challenges were how to find meaningful insights, to decide for the individual roles and responsibilities as well as the delicate balance of internal and external capabilities. | Section 4.4 Figure 5.4 | Feelings |
| | Two individual aspects that appeared to influence participants' decisions about technology use are:<br>• A tendency to participate in a shared endeavor;<br>• A feeling of co-presence | Section 4.4.1 Figure 5.1 Figure 5.4 | Choices Use and Feelings |
| | Retention of the co-presence and eventedness, involve to some degree retaining the volunteer's perspective. | Section 4.4.1 Figure 5.1 Figure 5.4 | Usage and Feelings |
| | One of the key affordances of various tools used throughout the CSC Program is its collaborative affordance. That is, the tools have properties that allow them to be used to collaborate. | Section 4.4.2 Figure 5.1 | Usage and approaches Choices |
| | The tools cannot be used for arriving at a precise decision. | Section 4.4.3 Figure 5.1 | Critical moment Feelings and assumptions |
| | The clarification of mutual roles and responsibilities is essential to effective utilization. | Section 4.4.3 Figure 5.4 | Feelings and assumptions |
| | Some participants mentioned that receiving informal peer support is an important alternative to receiving timely formal support. | Section 4.5.1 Section 4.5.2 Figure 5.3 | Support |

| | | |
|---|---|---|
| Participants equipped with various levels of expertise and areas of interests nourish the volunteering environment with ideas and knowledge that are befitted by volunteers based on their needs. Given the 'transformative' nature of such interactions (Pea, 1994), individuals acquire more expertise as the dialogue unfolds and they co-construct knowledge. There is also the opportunity provided by the online collaboration tools to go through a team evolution process and recognize common patterns in communication styles of group members. | Section 4.6.1 Figure 5.1 Figure 5.4 | Choices Usage & Feelings |
| Some participants expressed some issues of concern such as privacy. Still some of the participants indicated they would collaborate without technology, but they would prefer not to. | Section 4.6.3 Section 4.4.5 Figure 5.4 | Choices Usage and Feelings |
| The different tools enabled the volunteers to navigate through information, find personal routes and pathways. | Section 4.3 Figure 5.1 | Usage Choices Skills |
| The volunteers are also endowed with a flexibility that enables 'collaborative remix ability' (Boyd, 2007)– a transformative process which denotes the state of the information which can be recombined to further develop new concepts, ideas, and services. | Section 4.6.4.1 Section 4.6.4.2 Figure 5.1 | Usage and approaches Choices Skills |

94  Some Initial Reflections

At the heart of the CSC Program lies the process of project-based learning that enables the individuals to gain a shared understanding and construe a common basis for knowledge creation. This does not necessarily leave aside individual contributions and perspectives, yet volunteers are not required to segregate their work into discrete tasks to be completed individually and bring them together later on. Rather, they are required to make contributions to the point of views of their team mates for the mutual negotiation of meaning and the joint construction of a project by using online collaboration tools (Roschelle and Teasely, 1995). Coordination is a necessary element only when putting together the partial results of the discrete tasks of the related project (Roschelle and Teasely, 1995). On the other hand, the construction of a joint project through genuine collaboration necessitates a coordinated effort for a joint problem-solving (Roschelle and Teasely, 1995). It involves an interactive process that requires the participation of all group members for mutual negotiation and sharing of ideas (Roschelle and Teasely, 1995).

The review of evidence presented in the next figures has highlighted in more depth the online collaboration experiences of volunteers, considering the complex relationship between users (skills, knowledge, and assumptions), the tools and the volunteering contexts in which participants need to get involved.

Figure 5.1 displays the emerging theme 'Strategies' by providing information about the choices made by the CSC volunteers.

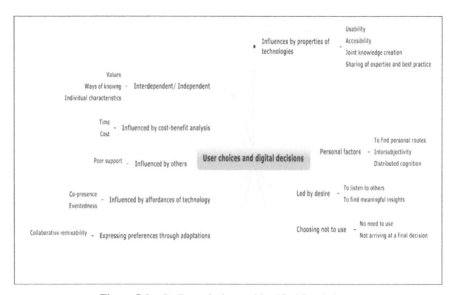

**Figure 5.1**  Coding sub-themes identified for choices.

Next, Figure 5.2 displays the emerging theme of experiences that relate to the use of online collaboration tools of CSC participants throughout their volunteering program.

Next, Figure 5.3 displays the enabling or inhibiting factors that emerge throughout the interviews with regard to the use of online collaboration tools of CSC participants.

Figure 5.4 displays the emerging theme of beliefs and intentions with regard to the use of online collaboration tools of CSC participants throughout their volunteering program.

According to Selwyn (2006), who came up firstly with the term "digital decisions", when individuals make empowered decisions to use or not to use technology, they exercise a genuine choice by taking into account its relevance, usefulness or even happiness caused by its usage throughout their everyday lives.

The choice for not using the technology is also evident in the CSC data. One of the underlying reasons for not using the technology was no being able to "get on with them". It is also evident from the data that several CSC volunteers think that it is up to them to take refined and complicated decisions for the usage of digital tools to aid their volunteering practice. The affordances and features of digital tools mainly underlie this decision-making process in addition to other factors.

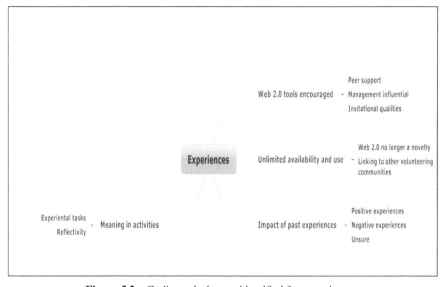

**Figure 5.2** Coding sub-themes identified for experiences.

96  *Some Initial Reflections*

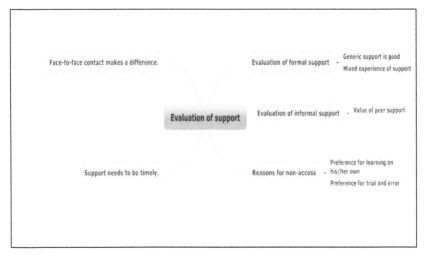

**Figure 5.3**  Coding of the sub-themes specified for the evaluation of support.

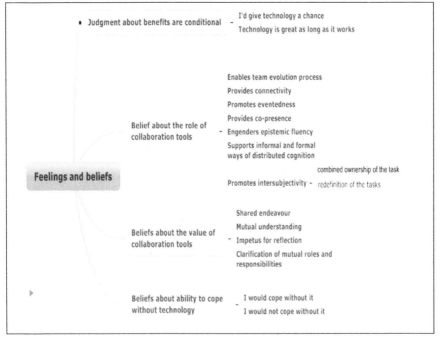

**Figure 5.4**  Coding sub-themes identified for feelings and beliefs.

The results from the study suggest that the opportunity of both being provided with a feeling of co-presence and eventedness are reasons why participants liked using online collaboration tools mostly (Section 4.6.3). CSC participants mentioned that the feeling of belonging to a networked community of colleagues sharing resources and asking for support the value of peer support is also an influential factor (Section 4.6.3).

Finally, if there were a particular amount of dependence on collaboration tools due to the assumption that it facilitates an easier collaboration; CSC participants preferred to refer to specific aspects and stated their views in a confident manner, rather than just being in favor of a particular tool or using it.

## 5.1 The Generalizability of the Results

By comparing the CSC Program to the wider population of employee volunteering programs; or by considering how representative CSC participants are in comparison to the wider organization's population it is possible to discuss about the empirical generalizability of the CSC Program results. I think that it would be more relevant to talk in terms of theoretical generalizability due to the qualitative nature of participant experience studies. As Green (1999) argues, theoretical generalizability corresponds to the degree to which the relevancy of the results reaches beyond the relevancy for the actual participants in the study. In a similar vein, Sim (1998) states that theoretical generalizability refers to the extent to which universal theoretical insights are to be offered by data for the purpose of projection to other contexts or situations (Sim, 1998). The results of this study do not only build on existing theories and discourses regarding the use of online collaboration tools, but also challenge the reader to rethink his or her understanding and application of these theories.

The CSC group interactions among volunteers were mediated by online collaboration tools and the volunteering practices (creating a shared meaning in the context of joint activity) were mediated through online artifacts (Stahl, 2001). The use of online collaboration tools for intersubjective meaning making has been visible through the CSC Program.

The terms of ""intersubjective learning" (Suthers, 2005) or" group cognition" (Stahl, 2001) might seem too vague for some readers. The problem of intersubjectivity is of particular relevance here as my aim was to understand how or whether learning is facilitated by collaborative interactions among the volunteers. Learning occurs as a result of creating meaning (Suthers, 2005) or establishing links among divergent meanings within the context of a joint activity (Hicks, 1996).

## 98  Some Initial Reflections

'Technology-enhanced volunteering' refers to the mixture of the social affordances of digital tools, with new informal learning agendas and priorities that provide the potential for cathartic shifts in employee volunteering practices. Brown and Duguid (2000) states that the balance between human interaction and computer interaction is an important aspect with regard to the effectiveness of an online community. I, therefore, assert that when it comes to planning activities for the practice of "Technology-enhanced volunteering" it is crucial to address at least four factors: The collaboration process, the infrastructure (e.g. the system), the motivation, and the resources/content (Figure 5.5).

In terms of the learning process, the CSC Program can be regarded as a non-formal learning activity as there is no predefined curriculum or structure for training. The volunteers could exchange ideas with their own team members or no-members who have useful contacts in community to set their own knowledge level accordingly. The different tools enabled the volunteers to navigate through information, find personal routes and pathways. The transactions of the volunteers aimed at facilitating, and validating understanding, and developing capabilities that will lead to further learning. Although the tools were not mainly designed for the purpose of learning, volunteers perceived its potential to support learning.

With regard to the motivational aspect, the following questions can be taken into account:

- Is the existing process of online collaboration caused or enhanced by the participant's own motivation?
- Does the motivation have a more external nature, i.e., volunteering task demands?

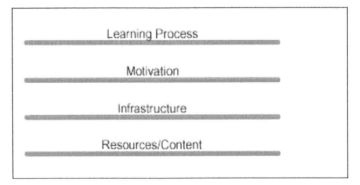

**Figure 5.5**  Model for online collaboration.

**Table 5.3** Questions for exploring online collaboration for 'Technology-enhanced Volunteering'

| |
|---|
| The Collaboration Process: |
| How is the collaboration structured? Is it e.g. formal and/or non-formal? |
| Which social affordances of the online collaboration tools used are of primary importance? (facilitating, and validating understanding, developing capabilities, increasing the level of connectedness, reflecting upon experience) |
| The motivation: |
| Does the motivation rely on external or internal factors? |
| To what degree is collaboration in itself motivating? |
| Is there a common ground established to convey the best practices through the means of these tools? |
| The infrastructure: |
| Which online collaboration tools are provided? |
| Are there any issues with regard to the ownership and control of the tools? |
| Are the tools 'context-specific' or imagined to transgress boundaries of the volunteering practice? |
| The resources/content: |
| Who controls the content/resources? |
| To what extent is 'collaborative remix ability supported? |
| Who describes the various responsibilities with regard to competence, accountability and copyright? |

I also created a set of more concrete questions (Table 5.3) with the aim to encourage reflection, as to support employees in developing an awareness of the tensions and potential perils with regard to the use of online collaboration tools for project-based learning practices.

## 5.2 Evaluation of the Methodology

The findings from this study highlight a set of elementary methodological issues with regard conducting a research and gaining an understanding of the volunteer perspective and in particular an understanding about the use of and impact of online collaboration tools on especially project-based learning activities.

My aim as a researcher was to implement a unique approach for the evaluation in order to convey real practices and experiences, therefore I adopted a two-part approach of an overall general survey to ensure that the volunteer's uses and perceptions of tools are matched with a set of in-depth case studies (through the interviews and digital artifacts).

I also developed an awareness about the methodological issues with regard to the use of specific digital artifacts such as personal blogs as a means to

## 100  Some Initial Reflections

convey the volunteers' experiences, therefore, I made the decision to make use interviews as means of conveying specific examples of practice. To offset the potential issues with regard to data collection via the online survey, I made use of standard semi-structured interview with various volunteers with the purpose of realizing adequate data collection and triangulation (Figures 5.6 and 5.7).

**Figure 5.6**   Overview of the new CSC platform.

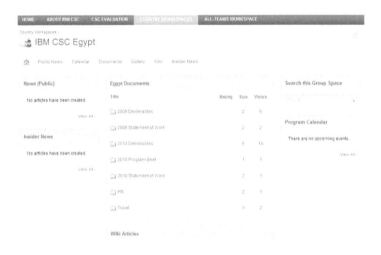

**Figure 5.7**   Menu items of the new CSC platform.

# 6

# Lessons Learnt

The CSC volunteering process relies on the proactive and dynamic collaborations of employees, whose work could further progress with the use of online collaboration tools.

The following key recommendations can be drawn out from the results that can inform the practice of all related stakeholders such as managers, trainers, IT professionals, social media experts, and other individuals interested in the practice of employee volunteering within organizations. Most of these recommendations are grounded in the confirmation of aspects that might cause difficulties for collaboration:

1. Raising awareness and acceptance for all interested parties interested in using online collaboration tools: As the volunteers prefer to switch among various technologies; they should possess a good understanding about the strengths and weaknesses of relevant digital tools in relation to their social affordances and influence on collaboration. As there is a range of refined and customized strategies for using technology to support collaboration increased awareness is crucial for effective utilization of the tools.
2. Increasing the level of grounding: 'Grounding' is important for the practice of 'Technology-enhanced volunteering, specifically when collaboration is occurring among volunteers across different countries. The volunteers should feel motivated to build a commonality by conveying the best practices through the means of these tools. Such a motivation would also constitute the ground for positioning the use of the digital tools a reciprocal activity and support the group to advance further in its phases of development.
3. Thinking more carefully about how collaborative remix ability can enhance the online collaboration of volunteers:Interactivity lies at the heart of modern employees' digital environment – ranging from actively looking for information on the Web to sharing ideas with colleagues, to

remixing information across multiple sources and media formats. Being networked learners, individuals are collaborating; sharing ideas and resources, exchanging draft reports, providing mutual support for each other. Online tools need to leverage the social affordances of technologies and it should be re-considered how this can enhance the support provided for volunteers. Also, shared repositories of content that aggregate relevant material will be a crucial cause not only for volunteers, but also for those keen to follow more individual learning paths.

4. Choosing technologies taking into account different tools made use of by the participants now and in the future: CSC participants displayed the ability to utilize various technologies and applications. The majority of volunteers were using instant messaging; and got involved in discussion forums; they also made use of internal Lotus Notes platform or popular Web 2.0 tools such as FaceBook, Ning, and uploaded visual artefacts such as audio or video onto the Internet. Most of the IBM volunteers access online learning materials via Edvisor (e-learning package developed for CSC Programme). In the future, given the shift in the digital ecosystem toward more individualisation and responsiveness online collaboration tools are required to be able to gather content in effective way based on volunteers' needs.

5. Designing and developing opportunities for employee volunteers to increase their motivation to utilize online collaboration tools: Two individual aspects that seemed to have an impact on participants' decisions about technology use are: a tendency to participate in a shared endeavour and a feeling of co-presence. The sociability aspects of these tools not only facilitate support for dialogical interaction; but also support for digital communities or relationships between individuals. By staying connected, they had the opportunity to validate their understanding, develop capabilities that will lead to further learning. Sustainment of the dialogue and requirements of the community play a more important role than personal needs.

A cautionary note should also be made, namely that social and organizational processes will play a role in terms of the implementation or rejection of specific technologies, both at the institutional level, and individual level. The lines may not be apparent for making a distinction between the transformational entities who would use these tools with the intention of making the workplace more open to new ideas, new collaboration techniques, and new goals for employees, and those conservative entities who consider the digital tools as a means of strengthening the existing processes rather than transforming it. This also underlines the need to be aware that these tools are not automatically "empowering" on their own.

Once a technology is adopted, further issues may arise, as expressed in the mentality "Now we've acquired it, let's use it". This argument is misleading as it emphasizes the use of computers to "support" employee volunteering activities, yet it provides no insights about how technology can support these tools. In such cases, there is the possibility that the adoption of a particular tool may lead to the disruption of an existing collaborative process, rather than its support. This can occur due to the shift of a collaborative activity which can be in good progress in an "offline" context onto an online environment, yet eventually users might be spending their efforts on details about the complicated digital tools despite the availability of "user-friendly" interfaces, and they would not even be able to reach the phase of making use of the tools to complete the collaborative work required. Therefore, the digital tools can act as a barrier to be avoided instead of acting as a means to support collaboration.

# 7
# Concluding Remarks

In agreement with LeBaron (2002), who suggested that the existence of technology is not independent from its usage, I believe that related activities, content, artifacts, and context with regard to any technology cannot specify new forms of practice on their own. These are indeed embodied within the practice (LeBaron, 2002). With the means of the organized actions of the individuals, an environment aids the desired form of practice. Collaboration tools and artifacts act merely as means in the way in which they are appropriated by their users in the directed practice of volunteering.

Technologies can provide many possibilities, but they cannot "fix" meanings (Suthers, 2005). Based on this fact, this research identified the specific benefits of online collaboration tools, and explored how their usage has been appropriated by employee volunteers for their practice of volunteering and how they influenced the process of their meaning-making. By doing so, it raised an awareness of the digital tools that provide collections of traits through which individuals can get involved in non-formal learning practices by having digital interactions with others.

It would be disingenuous and naïve of me to promise that the research study will by itself transform the online collaboration experiences of all users. I do, however, argue that the finding of book would increase an awareness amongst institutional stakeholders interested in the practice of employee volunteering to take further action and to provide direct responses to what the participants have said and done.

Collaborative efforts should start by focusing on the initial stage of awareness and receiving. CSC Senior Program Manager gave presentations about the benefits of using online collaboration tools throughout the volunteering practice. Because the volunteers were motivated to engage in activities that incorporated technology to improve their own collaboration, they were open to using electronic resources such as blogs, wikis, and Lotus Notes tools. IBM

Toolkit (https://www-146.ibm.com/corporateservicecorps/) has also provided valuable information about the sample blogs and wikis related to the CSC Program.

By this point, due to the perceived social affordances of the online collaboration tools, participants were able to move beyond the traditional boundaries of the volunteering practice. Throughout the process, the emphasis was on the establishment of a commonality by conveying the best practices through the means of these tools.

The study focused on employees' use and perceptions of online collaboration tools for their experience of employee volunteering. One of the possible follow-ups would be an in-depth study with a much wider scope across other employee volunteering projects or other non-formal learning practices which have a more representative basis with regard to the breadth of employees and their experiences. It would also be unique to conduct a further study which investigates the extent to which these methodological strategies conveyed volunteers' perspectives over a longer term. Besides, a comparative study with regard to the actual usage and the expected usage in terms of the CSC platform design and user expectations could also be conducted. Further information could be collected about learning approaches, online course design, and learning outcomes from the related stakeholders as well as through analysis of online content.

In the final analysis, the incorporation of online collaboration tools into the CSC Program is about change in the way the volunteers collaborate with each other, not about technology. This collaborative phenomenon raises the point about socio-technical systems thinking, which stipulates that technology in itself has little meaning. Within the context of employee volunteering, technology gains its value with regard to the collaborative interactions of the volunteers. It's about people and their behavior, not computers. It is about inventing new visions of employee volunteering in the context of a digital world. While the lack of online collaboration tools is a barrier to change, the presence of these tools does not guarantee change.

# Appendix A

## Informed Consent Form

March, 2010

This document is to confirm that we'd like to give consent to Ayse Kok for her research project concerning the use the social media contributing to the facilitation of organizational learning within the context of the partnership between IBM and the Canadian non-profit organisation Digital Opportunity Trust in the implementation of IBM's Corporate Service Corps program The research project will be conducted by Ayse Kok under the supervision of Dr Chris Davies, Professor and Course Director, Department of Educational Studies at the University of Oxford.

We have been informed by her that her project seeks to gain feedback from the related stakeholders such as academic staff and management on the effectiveness of social media toward facilitating organizational learning and online collaboration. The research project will include interviews and observations of participants. All interviews will be audio-taped and transcribed. Participants will have the opportunity to read relevant transcripts for accuracy.

Regards,

Kevin B Thompson
Senior Program Manager, Corporate Service Corps
IBM Corporation

# Appendix B

## Information Sheet

UNIVERSITY OF OXFORD
DEPARTMENT OF EDUCATION

15 Norham Gardens, Oxford OX2 6PY
Tel: +44(0)1865 274024, Fax: +44(0)1865 274027
general.enquiries@education.ox.ac.uk  www.education.ox.ac.uk

Director: Professor Anne Edwards

### *Information Sheet*

**Appendix B**

*Informed Consent for Interviews*

This form requests your consent to participate in a research project to gain feedback on the use of online collaboration tools throughout the CSC program in IBM. This research study - "The Effect of Online Collaboration Tools on Employee Volunteering: A Case Study of IBM's CSC Program"- investigates aims to show the significance of the use of online collaboration tools in supporting knowledge workers for the practice of organisational volunteering.

Project description: This research focuses on how online collaboration tools might be utilized by individuals in organizations for an organisational volunteering program. The research project is conducted by Ayse Kok- a MLitt student in University of Oxford- under the supervision of Dr Susan James in Department of Educational Studies at the University of Oxford.

Potential Benefits and Concerns: The project has the potential to assist with the improvements and new directions in terms of the use of online collaboration tools in similar future organisational initiatives.

Confidentiality: All information regarding this project will be kept confidential according to legal and ethical guidelines. All information associated with project participants will be kept only to the researcher herself. All identifying information will be cleansed prior to any dissemination of findings and any data disseminated will be in aggregate form. Every effort will be made to protect the accuracy and confidentiality of the data and of the respondents.

Participation is Voluntary: Your participation is entirely voluntary. You can freely withdraw from the project at any time without negative consequences and information related to you will be destroyed.

Results of the Study: The results of this study will be used to further develop and improve the use of collaborative learning technologies to enhance organizational volunteering and can hopefully used as a model for other organizations.

Questions? Please contact Ayse Kok (+90 532 360 47 65 or ayse.kok@kellogg.ox.ac.uk with any questions or concerns.

# Information Sheet

Please check the appropriate line to indicate that you have read and understand this letter:

\_\_\_\_\_ I give consent to participate in this project. I understand that I will participate in interviews designed to gain information about my perceptions of the use of online collaboration tools to facilitate organisational learning through our partnership with DOT.

\_\_\_\_\_ I would like more information before giving consent. Please call me at _____.

\_\_\_\_\_ I do not give consent to participate.

Signed: _____ _____
(Date)

NOTE: Informed consent MUST be documented by the use of a written consent form and signed by you or your legally authorized representative. Also, you WILL be given a copy of this form for your records.

# Appendix C

## Online Interview Questions

All information will be restored with confidentiality and only put into use for this particular research purpose.

**Description:** All questions will be designed to investigate the role of online collaboration tools in enhancing the practice of employee volunteering programs.

Interview questions

All of the interview questions are as stated below.

1. How does your organization make an effort to increase the usage of online collaboration tools during the CSC Program?
2. What are the main components in your company that facilitate the use of online collaboration tools within this context?
3. How many times on a daily basis do you make use of any of the online collaboration tools to exchange information with your colleagues and other related individuals involved in this CSC Program? Please give me some examples of what you use and how you use it.
4. Which tool does give you the best opportunity to collaborate with your colleagues?
5. Are there any downsides to using online collaboration tools for professional knowledge-building and -sharing? For example?
6. Do you think using technology – specifically for collaboration in this CSC Program can be improved? Please give specific examples.
7. What are you mainly concerned with regard to the use of online collaboration tools?
8. What are the factors that can increase your feeling of engagement with online collaboration tools?
9. What are the potential advantages of your exchange of idea via the use of online collaboration tools?
10. Is there anything else about your use of online collaboration tools that I could have asked you? Would you like to make any further statements?

# Appendix D

## Approval from Ethics Committee

SOCIAL SCIENCES & HUMANITIES
INTER-DIVISIONAL RESEARCH ETHICS COMMITTEE

Hayes House, 75 George Street, Oxford. OX1 2BQ
Tel: +44(0)1865 614871  Fax: +44(0)1865 614855
ethics@socsci.ox.ac.uk  www.socsci.ox.ac.uk

Co-ordinator of the IDREC
Social Sciences Divisional Office

Ref. SSD/2/3/IDREC

18th June, 2010

Ayse Kok,
Department of Education
15 Norham Gardens

Dear Ayse Kok,

Application Approval

Ref No.: SSD/CUREC1/10-410

Title: Demystifying Organizational Learning: A Case Study of IBM's Organizational Learning Program

The above application has been considered on behalf of the Social Sciences and Humanities Inter-divisional Research Ethics Committee (IDREC) in accordance with the procedures laid down by the University for ethical approval of all research involving human participants.

I am pleased to inform you that, on the basis of the information provided to the IDREC, the proposed research has been judged as meeting appropriate ethical standards, and accordingly approval has been granted.

Should there be any subsequent changes to the project, which raise ethical issues not covered in the original application, you should submit details to the IDREC for consideration.

Yours sincerely,

Chris Ballope

# Appendix E

## Overview of IBM CSC Program

IBM launched its employee volunteering program—called as CSCs—at the beginning of 2008. The CSC is an organizational initiative which was launched in 2008 and in which the IBM employees tackle the economic and societal issues of the less developed countries they have been sent to by especially integrating their technology perspective. The company sends its employees to 13 different countries. The scope of the volunteering program is dependent to some extent on IBM's size—recently, the program hosts under .2% of IBM's 400,000 employees per year. Yet, IBM started the CSCs as an indispensable part of a larger effort that provides leadership development opportunities for IBM employees. IBM aims to sustain the expansion of the CSCs geographically and to get as many participants as possible in the following years.

When selecting employees to participate in the CSC Program, IBM makes a global corporate announcement to encourage the application of its employees with different levels of experience. This bottom-up approach is also crucial for convincing the managers to support the involvement of employees and permit them to stay away from their regular jobs.

Because most CSC Programs involve a short period in the field, pre-assignment training is crucial to establish a cohesive team, equip participants with the related knowledge about the cultural and technical dimensions of their project and enable them to contribute to project activities.

IBM organizes a 3 months for preparation work regarding CSCs teams. The materials regarding to the preparatory work are conveyed and managed by utilizing a learning management system, and they involve both individual and team-based assignments. Weekly virtual meetings are held every week for 3 months before their departure for their work. Due to the fact that the team members are working in different IBM offices across the globe this preparation phase relies mainly on team-building. It also supports the employees in communication skills regarding their work across international teams. During the preparation phase other stakeholders that are included within the process are: team facilitators, country-specific advisors from the hosting CSC country, and alumni from prior CSC assignments in the same country.

The countries that IBM employees have been sent to within the context of this CSC Program are:

*Brazil*

CSC activity in Brazil started in 2009. The first two teams operated in Sao Paulo. In this emerging market, IBMers were collaborating with six different NGOs, addressing needs ranging from: education related to the community in under-served areas, diversity with regard to entrepreneurship, preventing violence, contribution to scientific education, digital inclusion, and digital entrepreneurship.

*China*

CSC activity in China started in 2009. The first two program locations are Chengdu, in Sichuan Province, and Shen Yang province. IBMers were working with different NGOs, chambers of commerce, governmental agencies and small business owners on topics ranging from logistics infrastructure to professional development.

*Egypt*

The first team of 10 IBM leaders were sent to Luxor, in Upper Egypt, in October 2009. The Egyptian Ministry of Trade and Investment was the main host entity and CSC assignments contributed to the government's agenda for developing internal trade and better integrating Luxor's informal markets; while also aiding national gender inclusion objectives. IBM CSC participants worked in partnership with national government and local stakeholders in Luxor to contribute from their own expertise in fields such as strategic planning, supply chain management, information technology, project management, and business development.

*Ghana*

The first CSC team traveled to the West African nation of Ghana. The goal of the project in Ghana was to improve business processes for targeted small and medium enterprises (SMEs) to the point where they can access financing for market expansion.

## India

Program operations started in 2009. Four teams were scheduled to depart to India in 2009. Two teams departed in September 2009 and two more in October 2009. Two were placed in Mumbai and the other two located in Ahmedabad. Teams were collaborating with government agencies, local NGOs and organizations of higher education targeting various business and technical limitations.

## Malaysia

The first CSC team of 10 IBM global leaders were hosted in Penang, Malaysia in May of 2008 and CSC teams continue to develop this Malaysian partnership going forward. Local partner institutions included the Universiti Sains Malaysia, D'Monte Child Care and Development and other local NGOs. CSC Malaysia teams worked directly with these local stakeholders in Penang to contribute from their own specialism in strategic planning, marketing, information technology, project management, and business plan development.

## Nigeria

Program operations started in 2009. Two teams served in Calabar, the capital of the Cross River State, and were scheduled to depart in August and October 2009, respectively. Teams were working with both non-profit and government agencies building local capacity and knowledge to drive future regional growth.

## Romania

The team going to Romania worked with the Romanian Center for Entrepreneurship and Executive Development to help develop employment opportunities in a global marketplace.

## South Africa

The CSC South Africa program started in 2009 with two teams destined for Nelspruit, capital of Mpumalanga Province. Projects are aligned with national agenda priorities of security and skills development. Local partners include the NGO Business Against Crime, the Mpumalanga Economic Growth Agency and the Mpumalanaga Tourism and Parks Agency.

## Tanzania

In Tanzania, the emerging leaders formulated development strategies and plans related to the Africa Wildlife Foundation, Tanzania Association of Tour Operators and Kick Start, a non-profit that develops markets for new technologies in Africa.

## The Philippines

In the Philippines, IBM team developed a marketing service facility and investment opportunity database for the Chamber of Commerce and its members, to provide the infrastructure enable a successful global marketplace.

## Turkey

In Turkey the IBM team worked with the Turkish Union of Chambers and Commodity to strengthen the governance practices of community organizations who are implementing a range of local initiatives that promote economic development and growth.

## Vietnam

The team assigned to Vietnam produced a plan for small and medium enterprises (SMEs) incorporating the needed market analysis and financial forecasting on behalf of the Danang Chamber of Commerce, in order to build strong community and economic ties for growth.

# Appendix F

## Individual Accounts

### Participant 1

| | |
|---|---|
| Title of Interviewee | Global Brand Manager |
| Date of Interview | 10 October 2010 |
| Age | 39 |
| Gender | Female |
| Home country | USA |
| Region | Africa |
| Tenure | 13 years |
| Job Family | Communications |
| Volunteering team | Sichuan Province, China (Completed) |
| Technologies used mostly for collaboration | • IBM communities (Ning), Facebook, Edvisor, Blogs, and E-mail |

*How does your organization make an effort to increase the usage of online collaboration tools within the context of CSC Program?*

Key for the success of our team during our 3 months preparation was the use of Ning. We chose Ning over an internal Lotus Connections Community because people outside of IBM could easily collaborate with us there. We used the site to share summaries of our meetings, post assignments (e.g., information we were learning about a country), share articles, and get logistical information.

We are also asked to chronicle our experience using new media tools such as blogs, video, and image collages. I've read about others' experiences, which helped me prepare and enables me to get to meet new colleagues and experience parts of the world I may never see. Also, I found writing a blog while I was on my assignment to be very 'cathartic'—it enabled me to pause and reflect on what was happening to me and how I wanted to be with what was coming next.

*What are the main components in your company that facilitate the use of online collaboration tools within this context?*

One of the great things about CSC that enables the participants experience to reach not only inside of IBM but outside as well is having a blog site. I've read about others' experiences, which helped me prepare and enables me to get to meet new colleagues and experience parts of the world I may never see. Also, the blog made me feel connected to friends and family back home.

*Please give me some examples of what you used and how you used it?*

Many members of the team and the previous China team are on Facebook and we did a lot of sharing there, especially photos. I personally posted over 12 different albums and nearly 1000 photos from my CSC experience.

Another technology tool that we use is something called Edvisor that enabled our team to complete an online learning program prior to our departure. While it was not collaborative in nature, the content that it provided for us supported collaboration.

And, there is a data repository, where all teams post their work products or deliverables from their assignments. I posted several things our team created. And, prior to leaving for my assignment, I searched the repository for other teams work that seemed related to what our team would be doing.

Also, recently, a Lotus Connections Community was established where CSC participants can share and explore information. I recently posted my CSC Presentation there.

*Do you think that a particular tool gives you the best opportunity to collaborate with your colleagues?*

We are a technology company and we create the kind tools you are researching so they are quite accessible. I think one of the challenges, however, is that IBM is very diverse with people who have different access to systems and different skill levels or fluency with technology. One of the things I've noticed is that there appears to be a transition period right now—some people who play with the new tools but there is not widespread activity or engagement. I think we are in the midst of a major culture change or shift.

Prior to my assignment, I went to the blog site about 2 times a week. While on assignment, I posted to my blog every other day. I used the data repository prior to and right after my assignment but don't envision using it again any time soon. I go to the online community about once every couple of weeks, if that—the community is more of a push: I wait for someone to invite me over.

*What are the factors that can increase your feeling of engagement with online collaboration tools?*

In my particular case, Ning and the IBM blog site were the most useful for sharing. I can see how the overall CSC Lotus Connections Community could be a more engaging space with more commitment from leadership and the participants to engage there. As I said above, I think the low usage now is a reflection of where the technology is culturally within IBM.

*Are there any downsides to using online collaboration tools for professional knowledge-building and -sharing? For example? Or any concerns you might have ...*

Not that I can think of at the moment ...

*Do you think that using these tools can be improved?*

11. I think using technology – specifically for knowledge-building and -sharing in this CSC Program can be improved. Key is that the program be run in a way that pulls CSC participants in. Perhaps the learning activities that are part of our Edvisor plan can be placed in the online community so people have to go into the community to access them; then, it might be natural to go to that place to share comments or ideas.

*What are the potential advantages of your exchange of idea via the use of online collaboration tools?*

I don't use them with any expectations, yet I must admit that I would love to see online collaboration become more of a way of working, versus something you do for social activity or when there's a 'collaboration' project.

*A way of working? Could you please elaborate more on this?*

I try to be a stimulus for change by sharing material; however, I tend not to go back if others don't engage. What's the secret to bringing an online community to life?

## Participant 2

| Title of Interviewee | Technical Sales Manager |
|---|---|
| Date of Interview | 15 September 2010 |
| Age | 38 |
| Gender | Male |
| Home country | USA |
| Region | Asia-Pacific |
| Tenure | 10 years |
| Job Family | Information Technology |
| Volunteering team | Luxor, Egypt (Newcomer) |
| Technologies used mostly for collaboration | Lotus Notes, Sametime voice suite tool, Skype, E-mail, and Ning |

*How does your organization make an effort to increase the usage of online collaboration tools during the CSC Program?*

Before I departed to CSC experience in Egypt I looked for all the communication tools available and authorized by IBM and it was very helpful while in the assignment. We had to have several calls with IBMers from all around the world to get the information our customer requested us and many of this meetings happened via telephone, Lotus Notes Sametime voice suite tool, Skype, and other tools.

*What are the main components in your company that facilitate the use of online collaboration tools within this context?*

I think that security is what most concerns the organization to allow the use of online tools. Some of these tools are very exposed to on line attacks to corrupt the system and this may cause loose of important information.

*So, you think rather than some enabling factors, there were mostly some disabling factors such as security issues?*

Yes, partially ...

*How many times on a daily basis do you make use of any of the online collaboration tools to exchange information with your colleagues and other related individuals involved in this CSC Program?*

I used online collaboration tools full time, especially the online chat tools like sametime or Skype. Our team facilitator encouraged us to use an online collaboration tool (Ning) in order to share photos, comments and discussions. Instructions were also provided on how to use an online project library (where we could store our completed project documents) and the CSC blog site (where we could share our experience with words and photos).

I would say Lotus Notes Sametime provides best opportunity for knowledge-sharing because you can chat, exchange files, call, conference etc. Key factors to enable the use of tools–access to internet, access to IBM network, easy to understand instructions on how to access tools. Key factors to disable the use of tools–download limitations for internet access when on assignment, limitations on Ning site that did not allow file sharing, difficulty in posting photos to CSC blog site (it was possible but required some complex steps).

*Which tool does give you the best opportunity for knowledge-sharing?*

The CSC Program library provided the best way to share knowledge. Although it is only able to be accessed via the IBM network, it provides a means of posting documents that can be categorized and tagged to enable for easy searching.

*Are there any downsides to using online collaboration tools for professional knowledge-building and -sharing? Or any concerns you might have?*

I don't think that there are any downsides to using these tools, we know talking face to face is the best option, but for a global team these tools are the best and always beneficial.

*Do you think that the use of these tools can be improved?*

I think we can always improve the use of these tools, but it's hard to tell how to improve, I think the tools can still get better, but for this the development team is the best to indicate how.

One factor that can increase my feeling of engagement with online collaboration tools is, certainly, the globalization, thinking globally ...We need to be connected to the world.

*So, can we say then that the feeling of connectedness is sort of what you expect from using these tools?*

Exactly

*Anything else you'd like to mention about the use of these tools?*

I think that security is what most concerns the organization to allow the use of online tools. Some of these tools are very exposed to online attacks to corrupt the system and this may cause loose of important information. We know talking face to face is the best option, but for a global team these tools are the best and always beneficial. We need to be connected to the world, but specially the possibility to be in contact with people from different geographies in a unique virtual tool and everybody receive the same information simultaneously.

## Participant 3

| Title of Interviewee | Chief Engineer |
| --- | --- |
| Date of Interview | 27 August 2010 |
| Age | 37 |
| Gender | Male |
| Home country | Brazil |
| Region | Eastern Europe |
| Tenure | 11 years |
| Job Family | Development Engineering |
| Volunteering team | Luxor, Egypt (Completed) |
| Technologies used mostly for collaboration | Lotus Notes, Wiki, E-mail, blog and Facebook |

*How does your organization make an effort to increase the usage of online collaboration tools during the CSC Program?*

At the preparation phase, we have had all the trainings on culture, security, and literature reading on social responsibility projects online so that our facilitators followed up our status every week. We also had the chance to have conference calls every week, we prepared and shared some learning topics during these calls.

*What are the main components in your company that facilitate the use of online collaboration tools within this context?*

I used the voice suite tool to connect with fellows from other IBM locations. After the assignment, I used Lotus Live tool to share my experience with future CSC in country assignees, as well as audio conference tool, once a week.

*What do you find most encouraging in terms of using these tools on such a frequent basis?*

During our assignment, the facilitator and the CSC management team encouraged us to use Facebook and Twitter to share our experiences. Our facilitator insisted that we write on the IBM CSC blog about our daily life and our assignment. We as a team noted down our experiences on the CSC Malaysia Team 4 blog.

Our organization has a lot of assets for collaborating ideas, our culture is also on knowledge-sharing. The CSC team all encourage us to use these tools and other means to share our experiences and knowledge.

*Which tool does give you the best opportunity to provide knowledge-sharing opportunities with your colleagues?*

During our assignments in order to share our documents; work products and deliverables we opened a team room. We uploaded our documents and shared our comments there within the team. Lotus Notes team rooms, **Lotus Sametime**-online messaging tool within IBM give us the best opportunity for knowledge-sharing with our colleagues.

*Any downsides or concerns you might want to emphasize with regard to the use of these tools?*

When working on assignment and internet access is not reliable or there are limitations on download size, the use of online tools can be difficult. This would especially be a problem if online tools were the only means the team had of sharing documents and knowledge. Ways to improve—the Ning site, as suggested by the team facilitator was not a useful tool, especially when most people are using Facebook. Would have been a better suggestion to build a facebook group for our CSC team and have people link to it.

*So, you don't have any concerns with using tools like Facebook for instance?*

We were very careful in commenting on Facebook and other websites about our assignments and life and all experiences since our clients and new friends were also reading those sites. Some of the cultural things we faced were very different to us and no comments were made on those since these are very sensitive issues.

*What are the factors that can increase your feeling of engagement with online collaboration tools?*

More communication brings more feedback and richer ideas. We benefited from our sharing ideas and documents by getting immediate feedback, more information at every stage. Ease of access, ease of use (clear instructions and guidelines for how to use them) make me feel more engaged with these tools.

*Are there any benefits you might expect from using these tools?*

Strangely enough, I found a "watched pot" effect in Egypt. I was counting on deep lessons from the program and those that are most evident to me now are actually from my blogging experience. I don't discount the program; I expect this finding is because I was grabbing program lessons throughout the month and as I look back I realize blogging was an element I minimized in importance until now.

1. I really enjoy crafting a story. It helps me put a pile of factors into a cohesive picture and articulate my view. In fact, this is something that energizes me at work as well, so the subject matters less than the process. (Note that in our client work, I found I excel at creating the logical and theoretical stories.)
2. I can't wait for a story to be final before posting. It turns out that a more complete story may actually muddle the message or worse, never get heard.
3. We each impact people without knowing it. As we shared our personal findings in the end of the program some people shared unexpected and significant changes in their outlook for their future and even beliefs. For every shared change I know there were many untold.

4. Sharing our vulnerabilities can help others. There can be risks with this for sure but when in the right context, opening up can be inspirational and encouraging to others while feeling darn good.
5. Lasting relationships are great; so are meaningful moments. I had some great moments with people I may never see or speak to again. When I consider it, I realize the relationships I maintain are based on the meaningful moments within them.

These aren't all new lessons, but they are renewed with different analogies and experiences. Relearning a lesson can be as important as learning it the first time.

## Participant 4

| | |
|---|---|
| Title of Interviewee | Regional Service Executive |
| Date of Interview | 30 September 2010 |
| Age | 42 |
| Gender | Male |
| Home country | USA |
| Region | Africa |
| Tenure | 15 years |
| Job Family | Sales |
| Volunteering team | Malatya, Turkey (Newcomer) |
| Technologies used mostly for collaboration | Lotus communities, Blog, Wiki and E-mail |

*How does your organization make an effort to increase the usage of online collaboration tools during the CSC Program?*

IBM makes an effort to increase the usage of online collaboration tools during the CSC Program in the following ways:
a. By sharing the information to the team through the use of collaboration tools.
b. By having discussions through the collaboration tools.
c. Tracking the experience.

*What are the main components in your company that facilitate the use of online collaboration tools within this context?*

Individual Accounts  125

The key factor in enabling my organization to facilitate the use of the use of online collaboration tools within this context is to bring together the team members. I use the tools once a day during the pre engagement period and mostly for building the layout of the workshop. Lotus Communities provides me the best opportunity to provide knowledge-sharing opportunities with my colleagues.

*Has there been a particular tool that you used mostly or found most useful in terms of exchanging knowledge with others?*

I think the blog tool was extremely efficient as it showed real time progress on our activities and also we could share our insights about the country, culture, people, and experience through it to several people who had its internet address. I would say the only downside would be the limited number of pictures size we could post to the blog. One of the great things about CSC that enables the participants experience to reach not only inside of IBM but outside as well is having a blog site. The CSC is a highly visible IBM program. CSC participants need to be aware of how to engage with the media, answer questions, and refer media to IBM's communication's organization.

*Are there any downsides to using online collaboration tools for professional knowledge-building and -sharing? For example?*

I think one downside to using online collaboration tools for professional knowledge-building and -sharing is that they cannot be used for arriving at a precision decision. It would be better if we had more collaboration during the post engagement experience.

Also, we should bear in mind that in some countries such as Turkey, you can't just rely on email to discuss issues or arrange meetings.
    F2f is still the major means of communication when it comes to meeting other stakeholders from partner institutions such as NGO's.

*That is an interesting point. So, you think for you f2f communication was still more preferred in comparison to other tools?*

Most of the time. For instance, whilst on assignment, the use of online collaboration tools was limited to the CSC blog site. I used this about once per week to post the narrative and photos of my experience and would review it about every two days to see what others had posted. Most of our file sharing was done via USB drives that were passed around the team. This enabled us to ensure we did not exceed our internet access download limit.

*What are the factors that can increase your feeling of engagement with online collaboration tools?*

I think the most important factors that can make me feel more engaged with online collaboration tools are:
 a. An engaging group
 b. A user-friendly tool

*Anything you would expect from the use of these tools?*

I would expect the following benefits in exchange for my contributions to the exchange of idea via the utilization of online collaboration tools:
 a. Increase team collaboration
 b. awareness among the team members

It would also be better if the number of people in the team that actively used the tool would increase.

*Thanks for your time ...*

## Participant 5

| | |
|---|---|
| Title of Interviewee | General Sector Marketing Manager |
| Date of interview | 19 September 2010 |
| Age | 37 |
| Gender | Male |
| Home country | Japan |
| Region | Eastern Europe |
| Tenure | 10 years |
| Job family | Marketing |
| Volunteering team | Mianyang, China (Completed) |
| Technologies used mostly for collaboration | Lotus Notes, Wiki, and E-mail |

*How does your organization make an effort to increase the usage of online collaboration tools during the CSC Program?*

In order to make communications effectively and efficiently, we should use online collaboration tools among various CSC teams I think. As a global project, CSC Program enables us to facilitate the use of online collaboration tools, such as e-mail, Skype, phone calls, messenger, etc. I think that these tools not only lower communication costs, but also narrow the distance between the volunteers in different countries across the world. So, developing an awareness of these tools is necessary.

*Please give me some examples of what you use and how you use it.*

I use the tools any time per day to communicate with related parties. Mostly, e-mail and Sametime due their ease of use and effectiveness.

"The use of the collaboration tools is unique in that it will provide for both IBM CSC volunteers and local communities to learn from each other and tell their stories about their building a Smarter Tianjin."

*Which tool does give you the best opportunity to provide knowledge-sharing opportunities with your colleagues?*

Capturing the CSC best practices via an online platform is really useful. For instance, include sample decks used on a kickoff conference call. We had one that included a slide from the Program Manager introducing various online tools. We also rotated the discussion forum leadership around the team, and had the person scheduled to lead the next topic be the scribe for the prior discussion on the wiki. That way they were sure to be dependent on contributing more, since it would affect the agenda of their discussion.

*Do you think using technology – specifically for collaboration in this CSC Program can be improved?*

I believe that the utilization of digital tools will increase in professional knowledge-building and -sharing. We can do more in social media with our NGO partners I think. We offer a social media platform for our employees to blog, exchange pictures and post videos. A small area is provided for each NGO partner to write an overall summary.

The crucial question to ask is what value could be established by doing more in this field. If we had an obvious answer to that question we'd invest more in this space.

I use e-mail constantly. I think e-mail gives the best opportunity to provide knowledge-sharing opportunities with my colleagues. If we had a transparent and simple online platform for sharing knowledge that would be helpful.

*What are you mainly concerned with regard to the use of online collaboration tools?*

Multi-tasking, privacy, decreasing focus on content are my concerns when it comes to using these tools for knowledge-sharing.

*What are the benefits that you expect in return from your contributions to the exchange of idea via the use of online collaboration tools?*

The use of the collaboration tools is unique in that you get to cycle through a team evolution process multiple times and experience patterns in team behavior/dynamics – and practice guiding and influencing people from different cultures. Very rewarding ...

Also, efficiency would also be a clear goal in terms of my expectations in return from my use of these tools.

*Anything else you would like to add?*

IBM takes skill-based volunteering very seriously. As IBM employees around the world help local communities during the CSC Program they also find the opportunity to use their various skills. They all share the same objective of using their expertise for being helpful for others. Their volunteering roles range from being a fire fighter to teacher or technician." CSC Participant [5]

The culture of IBM and the corporate values have an impact on everything we do. This is true for all regions across the globe in which we conduct our business. Especially, in the developing world we make investments for local communities by making use of our technology and expertise. Besides, IBM has a team of professionals who are interested in citizenship in a professional manner and who establish our local relationships in each country. IBM establishes a partnership with NGOs, educational institutions, governments and universities to increase our influence– with an emphasis on solving crucial issues at both a global and local scale.

In order to add value in service to local communities, there is an ongoing collaboration between IBM employees and non-profits via means of our On Demand Community, CSCs, or Smarter Cities Challenge teams.

## Participant 6

| Title of Interviewee | Software Distribution Manager |
|---|---|
| Date of interview | 29 September 2010 |
| Age | 39 |
| Gender | Male |
| Home country | Argentina |
| Region | Eastern Europe |
| Tenure | 15 years |
| Job Family | Human Resources |
| Volunteering team | Mersin Turkey (Completed) |
| Technologies used mostly for collaboration | Lotus Notes, Skype, and E-mail |

*How does IBM make an effort to increase the usage of online collaboration tools during the CSC Program?*

IBM does a lot of effort. We had many tools that we needed to use. I believe online collaborations tools were the correct way of preparing the project (2 months preparation before we left). We had online education modules for 2 months (prework before we leave), online library for references, online community for our team … and during the project, we used a lot the online blog on the IBM external website. I can say that IBM has focus a lot on online collaboration tools for this and it was great. Furthermore, the ability to get immediate feedback from others and to organize ideas have also been mentioned as critical factors. Again, the emphasis on individual utility is clear.

*What are the main components in your company that facilitate the use of online collaboration tools within this context?*

I would say that the keys factors is that already IBM employees are already familiar with online tools. IBMers used every day different online tools. These tools for the CSC program were just more tools to work with. As well, considering the fact that our team of 14 peoples were coming from 10 different countries, it was essential to have online collaboration tools. Of course we did have conference calls (this was essential too and needs to be held), but considering the time zones differences and the connexion line problem, online tools was better for everyone's convenience.

*Please give me some examples of what you use and how you use it.*

Before I left for my assignments, I would say that I use those online tools in average 3 times a day. Reading information (education), sharing ideas with colleges and just to see what new has been posted. During the assignments (for the IBM external blog), this as been maybe 1 or 2 times a week since I was already face to face with all my colleges. The blog was used to share my experience with other peoples (IBMers or not) outside of my team. Sametime and Lotus Connect is a wonderful program to communicate as long as you are need to get in touch only with IBMers ... but people outside – don't have this effort ... therefore, mailing was at the end the most used program during the CSC time. ... to share our documents we put it into Wiki at the end of the program – to have a library of our documents for the next team

In terms of challenges, it depends what type of learning profile (as an individual) you are. For me personally, I prefer face to face contact because I think less information is lost and as well the message is more well transmitted. But in this case this was not possible. However, I have to mention that conference calls are also essential. Conference calls are a complement in online tools. I think the online tools that were used were what could be done the best.

*Which tool does give you the best opportunity to provide knowledge-sharing opportunities with your colleagues?*

I enjoy using the wiki, yet it might be simplified in terms of its use. Navigating through it is not easy. A former CSC participant might say: "We've gathered the documents on this subject here", and then I won't be able to find which section it has been put into. I'd have to browse through each of the topics and open each of them to locate a particular document.

*Do you think using technology – specifically for collaboration in this CSC Program can be improved?*

Yes, definitely. I believe the tools interfaces could be improved. Sometimes, it was hard to find some information or to upload new data. The online tools could be more user friendly (easy to use).

For some participants, the focus should be on distinguishing between what is essential information and what is not.

*What are you mainly concerned with regard to the use of online collaboration tools?*

My key concern is to make sure that the correct message is well received by everyone. That the correct focus is placed on the correct information. When there is a lot of information (which was the case for our experience), it was sometimes hard to get to the essential.

Also, maybe that is not concern really, yet it is really important for me to have a sense that someone is speaking behind all those tools. I hope it can benefit to anyone in my team first (by sharing ideas, experience, etc.) but also to the large community in CSC program. I expect this to help other persons who will participate in the future to CSC program.

*What are the potential advantages of your exchange of idea via the use of online collaboration tools?*

No expectations really ...

*Ok, thank you very much for your time.*

## Participant 7

| Title of Interviewee | LA IGF Billing Team Leader |
|---|---|
| Date of interview | 26 September 2010 |
| Age | 39 |
| Gender | Male |
| Home country | US |
| Region | Eastern Europe |
| Tenure | 13 years |
| Job family | Finance |
| Volunteering team | Machakos, Kenya (Newcomer) |
| Technologies used mostly for collaboration | Lotus Notes, Wiki, and E-mail |

*How does your organization make an effort to increase the usage of online collaboration tools during the CSC Program?*

IBM is priming to pump in terms of initiating the use of these tools during our CSC practice. There's been a vast amount of information I couldn't access without any technology.

I think access, and other people using it (i.e., network effects) are the main aspects that enable IBM to facilitate the use of the use of online collaboration tools.

*How many times in a day do you use any of the online collaboration tools on average?*

It really depends, in general, I use these tools consistently in my day, I can really give no specific number.

*Which tool does give you the best opportunity to provide knowledge-sharing opportunities with your colleagues?*

Instant messaging gives me the best opportunity to provide knowledge-sharing opportunities with your colleagues.

*Are there any downsides to using online collaboration tools for professional knowledge-building and -sharing? For example?*

I don't see any downsides to using online collaboration tools for professional knowledge-building and -sharing. I think it is also about listening to other volunteers and let those folks know they've been heard. It doesn't mean we do everything others suggest, but they have to know that you listen and care.

The main issues relate to communication: when and how do we talk amongst ourselves? So you'd consider more leaders would put together communication plans and apply them.

*What are the factors that can increase your feeling of engagement with online collaboration tools?*

I think the team also shall tell you (and each other) how they prefer the information flow to occur.

Certainly they build rules about issues like when to answer the emails, participants to be copied on, so they can assume each other as being responsible.

Individual Accounts 133

If a team member could not participate in a conference call, for instance, they should be informed that they're responsible for reading the meeting minutes or at least try to find out what was discussed. If there is explicit commitment, it's easier to hold people responsible—it was their opinion.

*What are the potential advantages of your exchange of idea via the use of online collaboration tools?*

I don't expect anything in return ...

## Participant 8

| Title of Interviewee | Technical Support Manager |
| --- | --- |
| Date of interview | 29 September, 2010 |
| Age | 41 |
| Gender | Male |
| Home country | Australia |
| Region | Africa |
| Tenure | 2 years |
| Job family | Information Technology |
| Volunteering team | Malatya, Turkey (Completed) |
| Technologies used mostly for collaboration | Lotus Notes, Blog, Facebook, and Lotus Communities |

*How does your organization make an effort to increase the usage of online collaboration tools during the CSC Program?*

We were encouraged to use tools like Twitter, Flickr, Facebook, and blogs. CSC has its own blog but we were also allowed to use or own personal blogs as well.

*What are the main components in your company that facilitate the use of online collaboration tools within this context?*

I really can't think of any at the moment.

*Are there any downsides to using online collaboration tools for professional knowledge-building and -sharing? For example?*

Privacy is my main concern when it comes to collaborating via online tools. Less obvious is IBM's clearly defined social media policy which provides guidelines and allows us to use non-IBM tools to write about CSC. I think it touched on that already, but it's something that needs to be considered and that's a big part of IBMs social networking guidelines.

Also, the biggest down side with Lotus Communities – it's easy to assume that someone has seen or read an update that's posted into the community web site, when in fact that might now. As an example – I had a problem with my notebook connecting to the internal network for several days. During this time a few team mates thought I had seen a document that was posted when I had not. There were confusing conversations (not helped by language barriers) that were only clarified once it was clear that I had not seen the actual document. Eventually, we sorted it out.

*Please give me some examples of what you use and how you use it.*

In CSC, we were using Facebook daily. Many of us were blogging at least once per week. We also used normal email and internal IM to collaborate. Lotus Communities were used by some of the sub-teams to collaborate as well. For what it's worth I don't think anyone in our team was using Twitter.

As I said we used Facebook daily – I think the fact that every one of us already had a FB account made it easy for us to quickly find each other helped (apart from the Chinese team members – Facebook is blocked in China).

Facebook isn't an IBM tool so there's a lot we couldn't share. Lotus Communities is internal so we can share internal documents there and we did. The biggest down side – it's easy to assume that someone has seen or read an update that's posted into the community web site, when in fact that might now. As an example – I had a problem with my notebook connecting to the internal network for several days. During this time a few team mates thought I had seen a document that was posted when I had not. There were confusing conversations (not helped by language barriers) that were only clarified once it was clear that I had not seen the actual document.

*What are the factors that can increase your feeling with engagement with online collaboration tools?*

It is depending on the day. When we run the presentations or had any customer visit it was mainly late evening to plan a get together but if we have done our work with the IBM team or had to work with the interpreter we have send many mails and version to each other to share our new presentation or documents. All our documents need to be translated before we could go to our customers (because nobody speaks English).

*What are the potential advantages of your exchange of idea via the use of online collaboration tools?*

We had always the possibility to be online and could work very easy on-line and send our presentation to each other ... It was different when we worked outside this city (one week in the earthquake area) ... and we had to be happy if we have at least a light at night what was burning ... no Internet connection was available. I believe especially in developed countries you don't have always a good connection.

*Is there anything else about your use of online collaboration tools that I could have asked you? Or anything else you would like to add?*

Ohh, it just came to my mind ... I think these statements might be relevant for your prior questions.

One thing we talked about was a way to create some way for us to communicate that was neither on a public space (like Facebook) or attached to our IBM email. Some way for us to collaborate that was insulated for our 'day jobs' and also private so we could be more open than we can be on an external tool. E-mailing is still in my eyes the best tool if you want to reach one person ... more persons – to communicate – would be a kind of Facebook ... It helps to understand the diversity between each of us internal and external around the globe.

Security is an issue as well. I think it touched on that already but it's something that needs to be considered and that's a big part of IBMs social networking guidelines.

I don't know about any factors that make me feel engaged – I've thought about this a lot in relationship to Facebook and Twitter. I'm personally very engaged user of Facebook but I never really 'got' Twitter. I have an account and have had it for ages but rarely use it. Honestly I don't know why ... This 'Give/Take' relationship is crucial to me. I expect to get as much out of it as I put in. Maybe that's why I don't get Twitter, I don't put anything into it ... :-)

## 136    Appendix F

*Oh thanks a lot for these additional statements. So, you mostly used Facebook?*

There was another forum, yet its use was not simple. It did not give any notification about whether or not you had read the materials, so you had to just browse through to see if something new has been added. It really became complicated.

Also, since we had many tools to use, maybe a unique tool which combines all the functions of each online tools integrated in one unique tool. Many passwords to remember and many online tools to check.

*Ok, anything else?*

I spend a lot of time for volunteering and interact with and communicate with different people; I would not prefer this interaction to occur via the Internet since it does not offer the same quality level of communication.

## Participant 9

| Title of Interviewee | Senior Managing Consultant |
| --- | --- |
| Date of interview | 30 September 2010 |
| Age | 37 |
| Gender | Female |
| Home country | Poland |
| Region | Eastern Europe |
| Tenure | 8 years |
| Job Family | Consulting |
| Volunteering team | Chengdu, China (Completed) |
| Technologies used mostly for collaboration | Lotus Notes, Blog, Facebook, and Lotus Communities |

*How does your organization make an effort to increase the usage of online collaboration tools during the CSC Program?*

I have nothing against the use of technology support my learning, as long as it functions in a useful way. With the use of Sametime or IM, there is a potential for too many interruptions. The IBM blogs that we used can definitely be improved, make it more user-friendly. The important question to ask should be which kind of value to create by contributing more to this area. If there were an obvious answer to that question more could be invested into this space.

On the second day, once the program was finished, I realized that I started to miss my hectic days running between IBM, NGOs and the government of Chengdu. As my friends put it: "You cannot stand being lazy." In other words: I don't want to feel as if I can't be of use for anyone. Although during the last three weeks, I worked very much due to the project, my enthusiasm was at a highest level as I felt like providing some value for the team.

I remember my first day of fieldwork, while I went to the airport to meet one of the IBM's well-known engineers, Michael, who was a real gentleman; I felt like he maintained a distance although he was trying to act friendly toward me. Once we dropped his luggage at the hotel we had a very intellectual and pleasant conversation about different things such as food safety or Kessinger's book about China. One week later afterwards, the reputable engineers of IBM, along with the global corporate directors would gaze at my presentation about Dujiangyan Irrigation project and say: "You better remove the square disappear from the left . . . or you might prefer us to be out of sight in front of you." With the passage of time, we became so good friends that it was difficult for us to hold back tears.

One of the first things I learnt about is how important it is to get prepared. As my project was about cloud computing technologies I had a gamut of terminologies to grasp. Every day, I woke up earlier than others to get prepared for the project meetings. There were even jokes among us such as "I even did not work for SAT so hard." Yet, the preparation turned out to be worth it as I was feeling confident sitting in front of director.

Also, another important thing I learnt about is that Google Translate can never replace a good interpreter as there could be many cases in which the relevant and non-sensitive word needs to be selected. Of course, with regard to learning Chinese I have a long way to go, yet apart from the language I also learnt about interpersonal relationships, negotiation and coordination.

*What about your use of these tools? Do you think you can mention how and when you used them?*

I was asked to use blog and Wiki for the CSC Program.

Actually, I don't use the online collaboration tools too much for my CSC assignment. Maybe the following two points are the reasons . . . (i) In most of the time, I worked with my team and clients closely. That is to say, though I'm depends on email and phone call for my business, I'm not used to use online collaboration tools in my work. (ii) Only one of my sub CSC team was using the online collaboration tools—wiki—for his work (we have five persons in the sub team). At least, the guy who also gave up to use the tool.

We sit together around a table; share documents directly by email. Only to talk something "private", we will use IBM Sametime to exchange the information. I recalled ... I wrote my blog 4 times during the 4 weeks CSC Program.

E-mail provides the best opportunity to provide knowledge-sharing opportunities with my colleagues.

Our pre-work for the CSC assignment was performed through conference calls and setting up Lotus Connections Communities, in order to have international team meet prior to in-country assignment. IBM supports us for using many collaboration tools. Also, getting valuable insights and information from other fellow members using the tools, and discussion of improvement ideas for the program/project were other ways of getting support.

*Ok, these responses are really very insightful for me. Do you think that the use of these tools can be improved at all?*

I'm not sure whether it can be improved, however, it's not good enough now. The information shared in CSC Program is less than my expectation and the quality is not good enough. Be frank, I rarely research information on CSC repository after tried several times ... no high quality information I can have ....

*Any concerns you would like to mention regarding the use of these tools?*

My key concerns are:
1. Too much information. I can't know the quality of the information unless I digest it. It takes time.
2. Confidentiality.

*What are the potential advantages of your exchange of idea via the use of online collaboration tools?*

To have the right information and more information I want are what I expect in return from using these tools.

Even in other developing countries where the CSC Program is running you have so manyother problems (connection, missing/less computers, no skill to use a computer, don't understand English and, therefore, cannot work on the computer) that I don' t see a special knowledge-building on this. Example: Where ever we have been . . . we need Interpreters to understand our customers. Even financial statements couldn't be read by us. They write/speak in another language and we English . . . so even no computer/program can really help us here . . . if we used normal translation program . . . we still couldn't understand a word!

*What are the factors that can increase your feeling of engagement with online collaboration tools?*

You can send out daily tweets, update your Facebook or Ning page, and conduct webinars but those are one-way, broadcast media, at least the way most people use them. The emphasis is on how you better state what you would like to put forth and what type of channels would be suitable for conveying that message. It's this remixability that offers meaningful information and feedback that can be acted upon and form further interactions among volunteers.

*Anything else you would like to add?*

I don't expect any benefit but maybe people understand me better when I exchange my ideas and thoughts to the world – because what I learned in this CSC Program is . . . that culture is so different . . .. IBM internal and external . . . 10 different people from all over the world living for one month on the other-side of the world in a different country with a different culture/language and sign-body language . . . You need to be patient for so many things and to react different than in the western world. But at the end you understand a little bit and learned a lot from a different culture. BUT: This knowledge will be never given from any online collaboration tool. This skill you only get when you stand next to each other!

*Ok, thanks a lot for your time . . .*

## Participant 10

| Title of Interviewee | Chief Engineer |
|---|---|
| Date of interview | 10 September 2010 |
| Age | 35 |
| Gender | Female |
| Home country | US |
| Region | Eastern Europe |
| Tenure | 8 years |
| Job family | Development Engineering |
| Volunteering team | Kenya, Africa (Completed) |
| Technologies used mostly for collaboration | Lotus Notes, Lotus Communities, and Sametime |

*How does your organization make an effort to increase the usage of online collaboration tools during the CSC Program?*

Our pre-work for the CSC assignment was performed through conference calls and setting up Lotus Connections Communities, in order to have international team meet prior to in-country assignment. IBM supports us for using many collaboration tools.

*What are the factors that can increase your feeling of engagement with online collaboration tools?*

We all have laptop computers with software for **VPN** connection so that we can connect remotely to the IBM network from anywhere; the tools are also available and supported by the IBM helpdesk.
 During pre-work, we used the Lotus Connections community at least once a week. Also use Sametime a few times a week to communicate to other team members, and email more often.

*Please give me some examples of what you use and how you use it.*

Lotus Connections Database for knowledge transfer that stays in a repository. Sametime or email for more information sharing.

*Are there any downsides to using online collaboration tools for professional knowledge-building and -sharing? For example?*

I have no issues putting technology into use for helping me grow and go beyond my comfort zone, as long as that technology is ... meaningful and ... Of course, it's got to function. ... With the use of Sametime or IM, there is a potential for too many interruptions. The IBM blogs that we used can definitely be improved, make it more user-friendly.

*Do you think using technology – specifically for collaboration in this CSC Program can be improved?*

The IBM blogs that we used can definitely be improved, make it more user-friendly.

Ohh and it just came to my mind that a concern could be how confidential information gets handled if the appropriate levels of access are not used.

During pre-work we used the Lotus Connections community at least once a week. Also use Sametime a few times a week to communicate to other team members, and email more often. Lotus Connections Database for knowledge transfer that stays in a repository. Sametime or email for more information sharing. Regarding the decision to socialize, you are choosing what you prefer. Yet for collaboration, you have to take what is there.

*What are the potential advantages of your exchange of idea via the use of online collaboration tools?*

Just to level-set everyone. I'm here with 12 pretty amazing IBMers, 4 from US, 2 from India, 1 each from Germany, Hong Kong, Italy, Australia, Canada, South Africa. We are divided into 3 teams of 4, working on projects with different Government agencies. A Canadian NGO (DOT, http://www.dotrust.org/) is managing our relationships and our projects.

My project is with the Kenya Institute of Curriculum Development (KICD, http://www.kie.ac.ke/). We had our first presentation to the client today to clarify the scope of work and to determine next steps. It was unlike other meetings I attended so far throughout my career. The room was set up with a center table and other tables in a U-shape around that table. All of the guests (12 IBMers, plus our IBM CSC manager, and our NGO host) were seated at the center table. The staff from the Institute (about 25 people) sat around us at the outer tables. The meeting began with a prayer for the people assembled and for God's blessing on the good work we were doing. It was followed by introductions from everyone in the room, Kenyans first and then the IBM/NGO guests. When the introductions were complete, we broke for tea, which is served in china tea cups, with pastries. The tea break lasted about 15–20 min. When it was over, we got to work. Our team presented our work plan to the KICD and received a lot of important feedback from our hosts. The whole tone of the meeting was incredibly collegial, welcoming and filled with the promise of great things for Kenya.

So that point is something I have wanted to blog about since Sunday when we had our first meeting with our DOT host. Kenya has a new constitution, a new president and a plan for where they want to be in 2030 (Vision 2030). Over and over again, the Kenyans we meet demonstrate an incredible sense of optimism and energy for the future. Being from the US, it seems to me that this is how Thomas Jefferson and John Adams and their peers must have felt in the early days of our nation.

*Is there anything else about your use of online collaboration tools that I could have asked you? Or anything else you would like to add?*

Online collaboration tools are critical when working in global environments. And when there are big difference in time-zones then it is also important to use ones that do not expect quick answers back (like IM)... but that can be more asynchronous (like email, communities, etc.) Stimulating ideas, new insights and perspectives, synergies and something big being created from someone planting a small seed, engagement with others, building connections ….

## Participant 11

| | |
|---|---|
| Title of Interviewee | Finance Manager |
| Date of interview | 15 September 2010 |
| Age | 39 |
| Gender | Female |
| Home country | Chile |
| Region | Eastern Europe |
| Tenure | 10 years |
| Job family | Finance |
| Volunteering team | Aswan, Egypt (Newcomer) |
| Technologies used mostly for collaboration | Lotus Notes, Lotus Communities, Sametime, Facebook, and blogs |

*How does your organization make an effort to increase the usage of online collaboration tools during the CSC Program?*

12. IBM has many tools to use online collaboration in any time, not only for CSC. For the specific project in CSC they gave us USB connector to get wireless connection to internet and then trough AT&T get access to IBM tools for online collaboration like lotus notes, sametime, lotus communities, blogs, social media like Facebook, etc.

*What are the factors that can increase your feeling of engagement with online collaboration tools?*

The key factors that enable IBM to facilitate us the use of online collaboration are the equipments that they provide to us, like our thinkpad, USB, AT&T. Those three factors allow us be connected in everywhere at any time.

*Please give me some examples of what you use and how you use it.*

Blogs act as an informal space for conversation of thoughts and a historical data set in terms of project progress. I like and consider web logs as effective since they provide an opportunity for reaching out a bigger audience- that will often be specific to the person reading it. Blogs to get informed about the experiences of other CSC volunteers.

Almost all the time when we worked for presentation, we used to talk with other team members or other IBMers that could help us from their country and to contact local collaborators too.

*Are there any downsides to using online collaboration tools for professional knowledge-building and -sharing? For example?*

I didn't have any concern because we use secure tools ...

*What are the factors that can make increase your feeling of engagement with online collaboration tools?*

Easy to use is the key.

*Do you think using technology – specifically for collaboration in this CSC Program can be improved? Please give specific examples.*

For some projects, this occurs by mostly making use of the wiki to collaborate on voluntary projects. Perhaps the learning activities that are part of our

144  Appendix F

Edvisor plan can be placed in the online community so people have to go into the community to access them; then, it might be natural to go to that place to share comments or ideas.

*Is there anything else about your use of online collaboration tools that I could have asked you? Or anything else you would like to add?*

I don't expect benefits in return. Yet, I think we should also remind ourselves that mistakes should be treated as opportunities for development and learning. Don't get frustrated if the well-known social media mistakes occur. Acknowledge that problematic posts happening accidentally are also part of the process. Make use of such cases as learning opportunities to specify how you can deal with situations the next time they happen.

## Participant 12

| | |
|---|---|
| Title of Interviewee | Advisory Procurement Professional |
| Date of interview | 17 December 2010 |
| Age | 39 |
| Gender | Male |
| Home country | China |
| Region | Eastern Europe |
| Tenure | 10 years |
| Job family | Project Management |
| Volunteering team | Malatya, Turkey (Newcomer) |
| Technologies used mostly for collaboration | Lotus Communities, Sametime, Webex, and blogs |

*How does your organization make an effort to increase the usage of online collaboration tools during the CSC Program?*

During the pre-work period of the CSC Program, we used Lotus Notes Community to share the information with each other and Edvisor online learning system to have the training.

During the CSC Program in-country period, we used IBM Blog to share the CSC experiences, Sametime instant messaging tool to communicated with each other, and Facebook to share the photos and have fun.

*You seem to be positive in terms of your opinions regarding the use of the tools. So, my next questions would normally relate to how the use of the tools can be improved within the context of CSC program? Maybe, we can skip that question.*

Many positive things. It was the interaction among all other CSC participants. Their encouragement for making me try out new tools/applications ... I had a specific problem. Once I informed them about the problem and they suggested how to deal with, or they would simply state: "I am doing this", so then it gave me the impulse to try the same. If I was happy with the way it functioned then yes. I have used it. The positive aspect about this was I constantly learned new things. This enable me to develop the courage for trying new things and extending my space ...

Some improvement points still exist. For instance, the Edvisor courses were too long and did not match-up with the time estimates to complete; however, the books and travel tips were quite good. It seemed that contacting clients prior to in-country arrival seemed hit-and-miss, and were not easily facilitated ... maybe the Program Office could help facilitate some pre-arrival online interaction with the clients a little better, other than just supplying the contact information.

Also, sometimes, emails might take longer to write. Also, one has to explain and then wait for response. As even a few hours might be late for a reply in some cases there needs to be immediate support.

The management team of organization should encourage and support the use of online tools in order to build up an atmosphere and become a habit.

*How many times in a day do you make use of the online collaboration tools to exchange information with your colleagues and other related individuals involved in this CSC Program?*

Many times a day! as mentioned, Blog. Web Conference. I also used Facebook to share the fun; using Sametime to chart with each other; using Blog to publish the experience ... Making use of new tools, I wonder if I can benefit from it right now, and then decide upon whether I would need training about it. For me, it's better to spend an hour playing around with it on my own rather than a much longer training, as he course will have taken you quite a lot of time.

*Are there any downsides to using online collaboration tools for professional knowledge-building and -sharing? For example?*

One of the downsides of using online collaboration tools for professional knowledge-building and -sharing is that they cannot be used for arriving at a precise decision. Very often the discussions tend to be elaborating and not focused at arriving at a conclusion. Someone needs to be a moderator in the discussions and it takes time to find who could be the best person to do this when compared to a face-to-face group interaction.

*Which tool does give you the best opportunity to provide knowledge-sharing opportunities with your colleagues?*

During the CSC Program, our sub-team used web conference tool (Webex) to have meeting with the remote stakeholders.
    The audience may not be comfortable to use online tools to take knowledge-sharing, and they can not be concentrated. Quality of web conference in terms of internet speed.

*What are the factors that can increase your feeling of engagement with online collaboration tools?*

Save time for traveling; Easy to use. More audience can attend; Flexibility; Save time and money. Also, it is important I think to keep in mind for management: Highlight the volunteer's successes. Recognition is always appreciated. Share the results of the social media work done by your teams with other members of the organization during meetings and through organizational newsletters and emails. Focus on what's working and share that with others.

*What are the potential advantages of your exchange of idea via the use of online collaboration tools?*

Increased team collaboration and awareness among the team members are the benefits that I'd expect in exchange of my contributions via the use of online collaboration tools.

*Is there anything else about your use of online collaboration tools that I could have asked you? Or anything else you would like to add?*

An engaging group and a user-friendly tool such as a blog will make me feel definitely more engaged with online collaboration tools.

## Participant 13

| Title of Interviewee | Data and Storage Specialist |
|---|---|
| Date of interview | 19 December 2010 |
| Age | 39 |
| Gender | Male |
| Home country | Netherlands |
| Region | Asia Pacific |
| Tenure | 13 years |
| Job family | IT Architect |
| Volunteering team | Addis Abbaba, Ethiopia (Completed) |
| Technologies used mostly for collaboration | Lotus Communities, blogs |

*How does your organization make an effort to increase the usage of online collaboration tools during the CSC Program?*

I think that this travel can be described as a tour for listening. My CSC teammates and I interviewed several academicians, officials from the government and those working for non-profits as well as participants in the livestock economy. During my conversations with different people such a farmers or traders it was surprising for me to find out that no one had an idea about Market Information System of Livestock (LMIS) which was operating within the country for three years. The system aimed at developing the transparency of the markets. This made me realize an important lesson: Even though you may create something useful, people may still not know how to use it. I work for marketing, so, it was satisfying to discover how crucial that project can be.

During the first 3 weeks at CSC Program we really had a challenging learning experience. Apart from being exposed to the unique culture and friendly citizens of Ethiopia we were also offered the opportunity to visit the National Museum in Addis Ababa to the oldest human remains, from a few million years ago. It was from this cradle of humanity about hundred thousand years ago that human-beings started their journey which eventually included the whole globe. I think all of the human-beings have some roots in Ethiopia.

My CSC team had the objective to offer recommendations about the possible improvements for the LMIS. I was collaborating with other team mates from Australia and China on that project. We worked on recommendations with regard to the evaluation and analysis structure for the LMIS. This was one of the several projects being undertaken by the CSC team.

I would love to see to be a witness to a guy living in a village in Ethiopia and checking the prices of stock while interacting with others who offer a potential to buy these stocks by a mere click on his mobile phone. I think such technologies that put connectivity and transparency at the forefront could establish a butterfly effect. Within this field, any kind of service we as volunteers could offer to create that outcome will be time spent meaningfully.

*How many times per day do you make use of any of the online collaboration tools to exchange information with your colleagues and other related individuals involved in this CSC Program?*

I was using mostly blogs and online communities for my CSC work. Based on my observation I can say IBM is more inclined toward making use of different approaches to support bottom-up initiatives, especially in our CSR project where they're trying to make the most out of social media.

I don't think I appropriate its use depending on other collaborative activities, I use them in a supplementary manner i.e. looking for the wiki to find specific information and then asking my colleagues for it using e-mail.

*Which tool does give you the best opportunity to provide knowledge-sharing opportunities with your colleagues?*

In this age, we are using new technologies all the time, and introducing new ones to our way of working as new technologies appear, it's not a case of "appropriating"; we used to work using multiple methods, some conventional, some digital. In my opinion, the CSC team should agree for which technology tools to use for which purposes. A plan about the use of social media is only going to succeed under the condition of everyone's following it. We should be specifically be clear about keeping track of the meeting minutes, or recorded conference calls and other ways to stay up-to-date so that one does not lag behind anything.

*Do you think using technology – specifically for collaboration in this CSC Program can be improved? Please give specific examples.*

I think we need to have a more detailed grasp of the questions and approaches that will further feed into the bottom-up participation in social media. So, some thoughts regarding your questions:

- Ask volunteers what they would prefer their higher level management to support in this project. This question might appear as obvious, yet I think this one might get lost easily. If volunteers have made a step toward the use of social media, then they presumably have determined the factors that would make their use sustainable. Especially, they are likely to have determined the organizational obstacles that against their full use of social media. Give utmost attention to their statements with an open mind. Do your best to implement their approaches.
- Focus on eliminating obstacles with regard to the use of popular platforms such as Twitter or Facebook, rather than on creating new ones. I can say that volunteers interested in using social media have a gamut of ideas for making progress. The main issues seem to be related to the institutional obstacles put on their way. From a management perspective, it might be tempting to exert control over social media use, yet if they would like to support bottom-up projects, they better give up underestimating the value of digital collaboration tools, especially the social media. It might be better for them to collaborate with their team mates in terms of eliminating barriers and ensure that they are not creating new obstacles.
- Offer the employees resources and ideas that might support them in the implementation of their project. Keep track of articles and examples that might be shared with team members. This will provide a confirmation of an understanding of and support for what they are doing.

These are a few of my thoughts.

*Brilliant, many thanks for all of your insights. And what are the factors that can increase your feeling of engagement with online collaboration tools?*

Especially time-saving is such a central point. One should better think whether the use of that particular tool would be worth the time spending, for instance 1 h blogging?' So, do I spend one hour now and don't have to do it again, or do I continue my old habit – which takes about 10 min, yet shall I spend

10 min on a daily basis? In my opinion, often times it is worth going that extra mile to choose something that will turn out beneficial in the longer term. But it's a delicate balance – It's about dedicating one's time to start and do that.

*What are the potential advantages of your exchange of idea via the use of online collaboration tools?*

When I arrived in the country due my project a few months ago I came across an article on the plane about Gandhi which reminded me of one of his famous quotes about living as if one were to die tomorrow and learning as if one were to live forever. Keeping that in mind, I started my CSC assignment! Conveying my know-how to a highly visible national project that supports African communities was surely very satisfying, yet the professional know-how I gained was also very impressive I think. Collaborating within a developing country and gaining an understanding about its areas for development is a mall step on my learning journey. During the IBM CSC, an opportunity to make a difference in the world is provided to any participant. Such an amazing adventure for sure!

In order to improve the efficiency of the system me and my team-made use of a specific methodology of IBM that also provided support for the ministry to gain an understanding of the related technologies that would be helpful to manage the system on a daily basis.

Despite my several years of experience in the field of system design and delivery, I had to consider everything from a new perspective while working with the government officials in Ethiopia. Given the fact that even electricity is not accessible to many of its citizens subjects such as mobility gain an increasing strategic importance in such developing countries. In order to increase access and coverage the Ministry of Agriculture also was in favor of providing handheld devices for the farmers.

Moreover, this project did not only enable me to explore and take on new assignments and duties, but also to share knowledge and learn from other colleagues. To give a specific example, I had the chance to participate in IBM press conferences that would sound even unimaginable to me and I also organized meetings between the government officials among USA and Ethiopia under sponsorship of IBM.

Once I got back to my country I realized that I utilized my leadership skills while having been useful for a community in need. It is very satisfying to see the impact of our project. For instance, a new research lab has been opened in Africa which is an evidence for IBM's dedication to improving the lives of the human-beings in need.

Without any hesitation, I would suggest any employee to participate in this project. You not only stretch out yourself by getting out of your comfort zone, yet you also have a tremendous opportunity for making an impact.

Also, I think all these tools very handy – they are available for our immediate use without us changing our physical place to collect the same information. By talking to others about my volunteering project, I can gain a much better perspective on how my colleagues are viewing things. For instance, when working on project-related tasks, we each share our expertise yet also have trust in each other, keep track of each other's progress and try to keep each one up-to-date in terms of project related information (submission deadlines etc.).

## Participant 14

| | |
|---|---|
| Title of Interviewee | System Software Engineer |
| Date of interview | 15 January 2011 |
| Age | 39 |
| Gender | Male |
| Home country | India |
| Region | Eastern Europe |
| Tenure | 4 years |
| Job family | Development Engineering |
| Volunteering team | Aswan, Egypt (Completed) |
| Technologies used mostly for collaboration | Blogs, Lotus Connection |

*How does your organization make an effort to increase the usage of online collaboration tools during the CSC Program?*

IBM is a leader in collaborative technologies through many solutions and tools available from Lotus brand. We active use web 2.0 technologies such as online conferencing (Lotus Live), connections communities to collaborate with communities, profiles to identify skills/resources, activities for managing projects and more.

*How many times on a daily basis do you use any of the online collaboration tools to exchange information with your colleagues and other related individuals involved in this CSC Program? Please give me some examples of what you use and how you use it.*

## 152  Appendix F

A lot of technical issues such as troubleshooting with a blog about the CSC Program make up a typical day for me. Certainly I'm not the only one in facing these kinds of issues. Even the experts among us have found their job duties broadening given this tight economic conditions.

*Are there any downsides to using online collaboration tools for professional knowledge-building and -sharing? For example?*

There is no need to worry for us about site blocking as IBM encourages the use of various tools, so I was able to gain access and browse through the blogs, social networks, videos and forums that provided me with the responses I needed throughout the CSC Program. If required, I would also browse my own networks via social media.

*So, no downsides for you at all?*

Not at all ...

*What about your concerns?*

You wouldn't call it a concern exactly, yet I am not sure you are enough warning/informing the participants that they are responsible for making really use of these tools and that no one will do the efforts for them. We live in a world where individual is accustomed to have information digested and integrated in a system which is delivered ready to use. This is a lot about personal empowerment as well.

Unfortunately most of front line workers at the entities I collaborate with during the CSC Program, this would not have been the case. Many of these sites are blocked. If "blog" is in the title or URL, they are not able to locate it. If knowledge resides on a social network or forum, they are not able to visit it. Not even YouTube and online tutorials hosted there. Even major websites are blocked. Well also security could be one if enough precautions are not taken ... this is the new world and we MUST learn the rules of the game AND best practices.

*Do you think using technology – specifically for collaboration in this CSC Program can be improved? Please give specific examples.*

Not everyone is at the same level of learning and understanding of web 2.0 tools while self learning is vital, active use will provide motivation to those who are new to these technologies when they are working on the CSC Program. I am a web 2.0 advocate in IBM, we call us Blue Ambassadors. I personally have been making use of web 2.0 collaborative tools for over 3 years and my team in IBM Asia Pacific use it on a daily basis as this is our way to communicate with the team.

*What are the factors that can increase your feeling of engagement with online collaboration tools?*

When I did trainings and presentations throughout the CSC Program, volunteers will often ask our team how we "know so much." It's not difficult. There is no block on our access to the web, so when I pose a question, I can get a response. Receiving the related information makes me feel empowered. If you would like others to achieve their best at their work, the same level of access should be provided to them.

As I said before, there are not any downsides to using online collaboration tools for professional knowledge-building and -sharing I think. Naturally, there is learning all the functionality, but that's easy because you can do it as you go. Now, most tools are fairly intuitive. Right now, without widespread use and adoption, I guess one downside is that there is not much activity in response to posting something ... sometimes you can get more engagement by sending an email or setting up a conference call.

*Anything else you would like to add?*

I mostly use Lotus Connection because it is easy for me to bring order to my life, thoughts and work. If more volunteers were aware of what the advantages of using group collaboration tools such as Lotus Connection were I would also feel encouraged to use these tools. I can also use the discussion forums embedded within Lotus Connection for acquiring information and doing a comparison based on the rankings of search engines. On the other side, online forums often lead to more meaningful search results for specific knowledge that can be obtained internally than a much broader web search, and provide the added value that one can pose questions and get a human response.

## Participant 15

| | |
|---|---|
| Title of Interviewee | Retail Team Leader |
| Date of interview | 17 December 2010 |
| Age | 41 |
| Gender | Female |
| Home country | Brazil |
| Region | Asia Pacific |
| Tenure | 12 years |
| Job family | General Professional |
| Volunteering team | Sichuan, China (Completed) |
| Technologies used mostly for collaboration | Lotus Communities, blogs |

I review some of your questions. As I don't have really time I would like to provide some responses in general. Sorry about the last minute change, yet I am pretty busy nowadays.

*No problem at all, thanks for participating in the study. So, I look forward to hearing your general statements then.*

I think, one rich source of insights for your study might be trying to find out where individuals' passions are evident within their online conversations during the volunteering process. Some people including me prefer to express either positive or negative experiences in different host countries in their blogs.

I think some volunteers have a strong motivational aspect underlying their action. Passion, regardless of being negative or positive, might result in of emotional leverage. Since most useful marketing aims at a change in behavior, emphasis on a passion is the shortest way to more effective communication.

Also, such conversational tools like blogs offer other to understand the context of volunteers. The context offers the connections that help us grasp what goes from what people say to what people do.

For instance, digital conversations about something like burgers might vary from China to US, based on the participants of the conversations: e.g., moms might approach it from a different angel than foodies.

*That is a very nice example, thanks ... Anything else you would like to add:*

Individual Accounts   155

If we were to offer a random collection of facts or observations about our CSC Program in the digital realm, we need to go beyond surface to get in depth into the feelings and tension aspects in the conversation. And by gaining an awareness of the contextual who and where dimensions with regard to these expressed emotions, we can then make the connections to the pragmatic insights we all are trying to discover.

## Participant 16

| Title of Interviewee | Senior Consultant |
|---|---|
| Date of interview | 19 November 2010 |
| Age | 40 |
| Gender | Female |
| Home country | Australia |
| Region | Eastern Europe |
| Tenure | 12 years |
| Job family | Consulting |
| Volunteering team | Malatya, Turkey (Newcomer) |
| Technologies used mostly for collaboration | Lotus Communities, Sametime |

*How does your organization make an effort to increase the usage of online collaboration tools during the CSC Program?*

During our pre work period we used online tools for presenting and we also created a community to post our work and other important information so that everyone could log in and check what was new. Everyday during our pre-work. Examples (Lotus Community, Lotus Live meetings, SMILE) ... For Lotus Community and Live, one will need to sign up with their IBM user name and password and enter the community .... For Lotus Live, a conference id is provided that can be created and used. SMILE is used to make calls, there is a website where one needs to sign in and enter their number and the number they need to call, the operator will dial the number for you and connect you.

*What are the main components in your company that facilitate the use of online collaboration tools within this context?*

The fact is that social media, Twitter and all that, actually makes it easier for us to listen to our colleagues than ever before", That's right, people complain about distance and being physically far from others due to projects such as the CSC, but ask yourself: how many employees actually get unfiltered input from their others, even when they're in the same location? I think we did fine, we had all the information we needed on the Community and our facilitator also emailed us about new updates on the community. Encouraging everyone to use it and letting them know the importance of using the tool is also important.

*Which tool does give you the best opportunity to provide knowledge-sharing opportunities with your colleagues?*

I think the major benefits of these tools are:
- The ability to share stories online—The CSC Program utilizes the power of story-telling via an internal social media platform. They need to go further, but this is a good start.
- Clear connections to the business and other CSC stakeholders—Enough said!

With most of us participating in social media, these tools facilitate the infusion of streams of knowledge for collaboration at an uncommon level.

*Do you think using technology – specifically for collaboration in this CSC Program can be improved? Please give specific examples.*

I think, we as IBMers also evolved from impromptu experimentation to facilitated exploration in terms of the use of the social media. Yet, we need to determine some more challenges such as:
- Being equipped with skills for data management in order to make meaning of it and gain insights
- Deciding among the individual roles. We need to specify the delicate balance of internal and external resources and capabilities.
- Building a framework to embed social intelligence into every participant country within the CSC Program, take timely action accordingly, and then evaluate the impact of those actions.

*Very nice illustration, could you elaborate more on this please?*

A new genre of Web 2.0 tools are also evolving to contribute to overcome these challenges. They are developed to offer intelligence aligned with project aims, and to meet the challenges above by supporting the infusion of social intelligence across the project.

The teams that meet these challenges will add extraordinary business value by providing insight, better richer service, improved compliance, and much more. We should refer to main challenges as we move toward a better social enterprise.

Firms adding value to their districts have amazing experiences to share – experiences that can provide inspiration for others, encourage employees and increase their reputation.

It is saddening that most of the time these stories are never told. Instead, they are restored somewhere else and got forgotten amidst the daily chaos. While institutional policies that put barriers in front of communication among employees shall take most of the blame, many firms either don't have an idea of how to facilitate the sharing of their stories among their employees or they simply think that there is no time or resources to do it.

*Which tool does give you the best opportunity to provide knowledge-sharing opportunities with your colleagues?*

In fact, the opportunity for enterprises to publish the story of their community engagement has never been better thanks to social networking tools like Facebook, Twitter, and LinkedIn. Corporates, employees and even customers can easily share stories with regard to their eloquent community engagement via various places. Although it might be difficult to launch a blog, it is in fact not very difficult. The truth is that is exactly what other corporate volunteering programs have as a requirement all along.

*What are the factors that increase your feeling of engagement with online collaboration tools?*

The idea of using social media to listen seems counter-intuitive. After all, we get told constantly how easy it is to broadcast your message. You can send out daily tweets, update your Facebook or Ning page, and conduct webinars but those are one-way, broadcast media, at least the way most people use them . . .. Easy to use, information is found by date so one knows what has been posted recently. One common repository for all information.

158  *Appendix F*

Social media offers an amazing opportunity to receive input from our team members and make them feel being listened to and heard.

Its use gains more meaning by not displaying responses for all the world to see, yet getting necessary action based on that input.

*Anything else you would like to add?*

No, thank you.

## Participant 17

| Title of Interviewee | Senior Marketing Associate |
|---|---|
| Date of interview | 11 January 2011 |
| Age | 40 |
| Gender | Female |
| Home country | China |
| Region | Eastern Europe |
| Tenure | 10 years |
| Job Family | Marketing |
| Volunteering team | Aswan, Egypt (Newcomer) |
| Technologies used mostly for collaboration | Tele-conferencing. Lotus communities |

*How does your organization make an effort to increase the usage of online collaboration tools during the CSC Program?*

I think the CSC team should make a decision with regard to which technology tools they will utilize for which objective. Throughout these conversations, some concerns with regard to effectiveness or required training might be expressed. A plan about the use of social media is only going to be pragmatic if everyone follows it. I think once the plan is followed, we don't need to be afraid of mentoring those who roam away.

*Do you think using technology – specifically for collaboration in this CSC program can be improved? Please give specific examples.*

If the aim is to have one teleconference per week, discover why individuals are organizing other activities at that time and then ensure that participation is important. If they're sharing questions and answers to the forums on Lotus Communities, I express my gratitude and try to make others feel encouraged to participate. If they're not interested, one should ask them about the underlying reasons. In this way, one can find out whether the teammates are reliable or not.

*Are there any downsides to using online collaboration tools for professional knowledge-building and -sharing? For example? Or any concerns?*

We should keep our focus on the plan. A successful communication plan is not for being put into place once and being referred to. If team members are shown a role model or a leader they will be eager to participate in an ongoing communication.

*Do you think using technology – specifically for collaboration in this CSC program can be improved? Please give specific examples.*

Given the remote team communication, one better plans his communication and communicates his plan.
   The listening part is really underestimated. The focus is what to say, how to say it and which medium is the most effective for getting out that message. But both learning and leading require us to listen, too.

*What are the potential advantages of your exchange of idea via the use of online collaboration tools?*

It's the listening that gives you pragmatic information and feedback that can support you forming your interactions. It gives you an idea about the worries of the stakeholders or the potential value that might be added to them. Listening supports one in realizing what's happening with people and offers the best opportunities for providing value.

*What are the factors that can increase the feeling of engagement with online collaboration tools?*

I think, our CSC Program would not be able to progress well without an emphasis on its use of technology, especially the social media tools.

*Is there anything else about your use of online collaboration tools that I could have asked you? Or anything else you would like to add?*

I think the following volunteering behaviors set this CSC Program really apart from others:

- We invest more in social media.
- There is a bigger chance for us to receive support from the leadership. Social media is being championed by our leadership.
- We are also more likely to diversify our tools. We are more likely to use apart from Lotus Communities 'the big four' (Twitter, Facebook, LinkedIn, and YouTube). These tools are also important to our social-media efforts.
- We are also more likely to listen. It is about engaging, providing feedback and listening that involve a more open relationship with others.

Of course, it should be mentioned that all of this is historical. Think for a while what social media looked like a few years ago. We are never going back to that world. I expect that the core lessons will endure. We like to do new experiments, and we keep doing what works.

# Appendix G

## Online Survey

Survey: CSC Team Survey (14 questions)

> This survey will be used to assess the use of existing online collaboration tools such as IBM Web 2.0 tools and Lotus Notes including e-mail throughout the CSCs program.
>
> Book—"The Use of Online Collaboration Tools for Corporate Volunteering: A Case Study of IBM's CSC Program"—investigates aims to show the significance of the use of online collaboration tools in supporting knowledge workers for the practice of organizational volunteering. This research focuses on how online collaboration tools might be utilized by individuals in organizations for an employee volunteering program. The research study is conducted by Ayse Kok—a research degree student in University of Oxford-under the supervision of Dr Susan James in Department of Educational Studies at the University of Oxford.

You can respond anonymously to this survey. All responses will be held anonymously. Thank you all for the participation!

1. Which of the following tools or methods do you prefer to use to share knowledge, experiences or best practices within CSC team? (Required/choose at least one)
   IBM Web 2.0 Applications (weblogs, wikis, and social networking)
   - Lotus Sametime
   - Lotus Connection
   - Lotus Notes
   - Team meetings
   - Other _____
   
   Colleague(s) (Mentor/Buddy/Mentee ...)

2. Which of the following tools or methods do you prefer to use to share knowledge, experiences or best practices related to volunteering with the CSC alumni? (Required/choose at least one)
    - Colleague(s) (Mentor/Buddy/Mentee ...)
    - IBM Web 2.0 Tools (wikis, blogs and social networking)
    - Lotus Sametime
    - Lotus Connection
    - Lotus Notes
    - Team meetings
    - Other_____
3. According your point of view what is the most practical way of sharing best CSC practices or experiences? (Required/choose one)
   _____
4. How motivated do you feel to share your knowledge, experience and best practices related to CSC program? (Required/select one)
    - Feel not motivated at all.
    - Feel a bit unmotivated.
    - Feel motivated enough.
    - Feel motivated more than enough.
    - Feel very motivated.
5. How crucial do you think is the sharing of best practices, experiences, contacts, and knowledge overall? (Required/choose one)
    - It is not really important
    - There are some things that I can learn from
    - No idea
    - I use it while I can also learn from others
    - It's very crucial for me to make use of the experience of other team members
    - Other_____
6. How crucial is it for you to share best practices, knowledge and skill with your CSC team members? (Required/choose one)
    - It is not really important.
    - It might be sometimes important.
    - It is often important.
    - Always important.

7. What do you think about being supported in terms of providing enough information to all your CSC team members? (Required/choose one)
    - Do not feel supported at all.
    - Do not feel supported enough.
    - Do feel supported enough.
    - Do feel supported very much.
8. What do you think about your ability to get sufficient information from all your CSC team members? (Required/choose one)
    - I can't receive enough information.
    - The amount of information I receive should be improved.
    - The amount of information I receive.
    - I can receive enough information.
9. In case you consider you cannot provide, share and acquire best practices, knowledge and leverage the experience of CSC team members – what is/are the prohibition/s? (Required/choose at least one)
    -----------
10. How do you receive management support for sharing CSC related best practices, experiences and knowledge? (Required/choose at least one)
    -----------

# Appendix H

## Participant Responses to the Online Survey Regarding New CSC Platform Design

| |
|---|
| I like the simpler pages, because they are far easier to negotiate your way around, and are aesthetically more pleasing. Though, it is important to include all the necessary information. |
| I like C1 as a browser page as there is not a lot of information which can distract the attention of the user, which can cause an element of time-wasting. However, I do not think this should be the initial introduction to the CSC site but that there should be a link from the Welcome page to this. Results page B2 gives the user a clear selection of results but also provides a little further information without being overwhelming. The different colors of A3 give indications of content, alternative and additional information. |
| An opening like C5 is great simple and very nice and the next page should look like B4. The top menu matches the colors of the opening page and is very informative and colorful and such is very attractive. |
| A5 looks the most readable and is clear in its' layout – not too cluttered. I really like C5 but it doesn't really give enough information for the welcome page. |
| B1 – less busy and squashed than a1. c1 does not have a welcome page that explains what the site is about. I think if I found a link to the CSC page and it looked like c1 I would be put off by the lack of info of what it was about. b2 – The main text is much more spaced out and there is less on the page. But if we are going down the road of having less stuff on the page then I don't think that the latest additings and news at the right top of the page is really necessary. c3 – I like the drop down boxes being on the left top. to compare with b3, b3 has about 10 icons on the left top. Are these all needed. Because a lot of them are missing on c3, which leave me to believe that they are not essential. I don't like all the different colors on a3, as if someone needs a blue background, they will only be able to read the bit with the blue background, and if someone needs a pink background, they will only be able to read the bit with the pink background. |

| |
|---|
| C1 is good, but may not convey enough info. Perhaps a short paragraph at the top would help. |
| As long as there is a home button and main navigation at the top and the search is on the left and tips I am happy. |
| I did not like C1 because I like lots of information and in C1 I was expected to know what it was all about. I like to see menus on the left it feels natural. I like a2 because you have the one liner but you also have the small description. A3 is best because it makes the page more interesting to look at and it is always useful to have tips that you are not searching for! |
| I like both the 'a' and 'b' groups for layout. The problem with 'c' is mainly the welcome page. While it is fresh and uncluttered, it tells people nothing about the project. I chose most layout options in 'b' because I like the navigation bars at the top. |

# Appendix I

## The Influence of Volunteer Responses on Design of Stage Two Interview Questions

| Proposed Research Questions | Influencing Participant Responses | Emergent Interview Questions |
|---|---|---|
| How are collaborative learning tools used for your volunteering practice? | This question received the highest ranking which was an indicator that it was a crucial question to be included in Stage Two interview. The question has later on been revised to include a clear definition of 'collaborative learning tools'. | How does your organization make an effort to increase the usage of online collaboration tools during the CSC Program? |
|  | Volunteers' statements such as: "I can't think of where to start with this as there are many tools we use" prompted me to state the original question differently by focusing on the particular tools that the volunteers use specifically. | Which tool does give you the best opportunity to provide knowledge-sharing opportunities with your colleagues? |
| What are your beliefs about the benefits and challenges in using these tools for such a practice? | This question received the second highest ranking among volunteers, highlighting that it would be another crucial question to be included in Stage | What are the factors that can increase your feeling of engagement with online collaboration tools? |

| | | |
|---|---|---|
| | Two interview. One comment was "It may be useful to look at why we use these tools at all?" | |
| | The following participant comments suggested that valuable information could be put forth if participants in Stage Two were prompted to elaborate further on their challenges:<br><br>"I think time is one of the most important factor is the amount you can get out of technology." | Do you think using technology – specifically for knowledge-building and -sharing in the CSC Program can be improved? Please give specific examples. |
| | Volunteers' comments such as: "I can't decide where to make the start" prompted me to focus on the negative aspects of using the tools as well. | Are there any downsides to using online collaboration tools for professional knowledge-building and -sharing? For example? |
| No question | Additional statements such as "Also if more digital learning is favored it would have to be supported by adequate training I think." have also been taken into account. | Is there anything else about your use of online collaboration tools that I could have asked you? Or anything else you would like to add? |

# Appendix J

## Phase One Recruitment Email – Pre-Consultation

Dear Sir/Madam

My name is Ayse and I am doing my graduate studies in the Educational Studies Department in the University of Oxford in UK. Worked previously as a project manager for one of IBM's CSC implementation partners—UNDP office in Turkey—I'd like to research about whether and how corporate volunteers are using online collaboration tools and related technologies as part of the CSC program.

I am looking to recruit volunteers who have already completed their volunteering practice.

For the purposes of the project I put in use the following three definitions:

- Online collaboration tools refer to web-based technologies such as popular Web 2.0 applications like blogs, wikis, and more conventional web-based applications such as instant messenger, discussion boards, online chats, and e-mail utilized by several individuals with the aim of accomplishing a common task.
- Digital learning is any digital learning resource or activity that are made accessible through the means of the Internet. Materials put onto the corporate learning environment along with the use of wikis, blogs, discussion forums, e-mail, podcasts, or digital library applications can be cited as popular examples of digital learning.
- Social networking denotes the usage of a dedicated web site to interact or socialize with other individuals to exchange ideas, pictures, articles, and videos.

The project has two phases: the first phase involves finding out from volunteers what questions they think it is important for me to ask, as well as their ideas regarding the different methods that we could use to capture the volunteer "voice". The second phase involves contributing your experiences of the technologies defined above, so that a case study outlining these experiences can be developed.

I am hoping that you can spare 10–15 min to help me with Phase One of our study. If you would identify yourself as a participant in this study I would like you to do one or all of the following three things.

First, have a look at my proposed research questions attached and have a think about how relevant or important you think they are and by replying to this email let me know which if any of the questions you would delete or amend as well as any questions of your own you would add to the list.

Second, have a look at my proposed methods for capturing the volunteer voice and experience and by replying to this email let us know of any methods you would delete or amend as well as any method of your own you would add to the list.

Third, if you would like to take part in Phase Two of the study, let me know by replying to this email and I will send you more information. Just to clarify, responding to my first two questions will not be taken as indicating a desire to take part in Phase Two, unless you specifically indicate in the text of your email reply that you want to.

All your replies to this email will be anonymized and handled in the strictest of confidence.

Finally, as a small gift for the time you dedicated we have prepared a list of resources relating to accessing or benefiting from social media and related technologies for companies, which you might find helpful. You can access it at:

URL provided

Many thanks in advance for your time.

Kind regards

# Appendix K

## Example(s) of Artifacts Provided by Volunteers

### A Blog

Example(s) of Artifacts Provided 171

## Audio Files

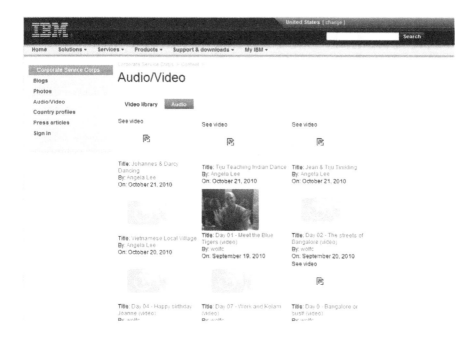

# Appendix L

## Screenshots Used to Assess Participant Preferences for the New Online Collaboration Platform

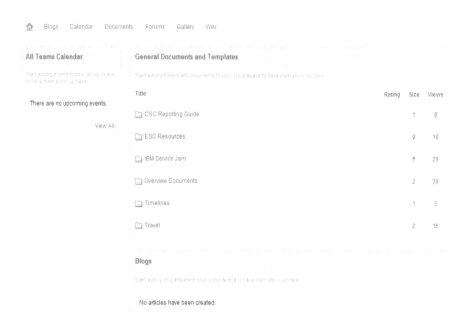

## Screenshots Used to Assess Participant Preferences 173

About IBM CSC

The Team

Where We Work

Partners

### About IBM CSC

Corporate Service Corps is a skills-based corporate volunteerism program launched by IBM in 2007 which is an employee leadership initiative. It is intended to help enhance global economic and social development and build the leadership skills of IBM employees as lobal citizens. The CSC program focuses on several priority issues: 1) Economic Development and Innovation, 2) Access to ICT, 3) Raising Global Standards in Education, 4) Broadening Cultural Awareness and; 5) Promoting Openness and Transparency, among others.

Digital Opportunity Trust (DOT) has been selected by IBM as one of the three global implementation partners for the CSC program

Corporate Service Corps program gathers teams of IBM Leaders with a diversity of skills, drawn from different countries and business units and place them in emerging markets to tackle important social and economic issues in collaboration with NGO partners from around the world

The IBM Leaders work on projects of significant value in developing countries, in four-week assignments. These teams will tackle real societal, educational and economic challenges, while at the same time experiencing a diverse cultural perspective and enhancing their skills

# Glossary

**Blog:** Web-based diaries where Internet users can write about their thoughts and provide information and links on various topics of their interest.

**Chat:** The process of being in communication with other individuals connected to the Internet in real time.

**Collaboration:** An interaction model in which authority and acceptance of responsibility among are shared to some extent among group members for a common purpose.

**Collaborative learning environment:** A pedagogical approach that includes various individuals coming together to work on a task, seek a solution to an issue and create a product by making use of web-based technologies to fulfill the synchronous and asynchronous requirements of these individuals.

**Collective learning:** Learning type in which the individual employee, as a result of the learning activity, was not only expected to be able to develop or change his competence, but eventually also to modify current work patterns and/or develop new patterns.

**Communities of practice:** A group of individuals who share common interest in a particular domain or area, and aim at increasing their own knowledge through the process of sharing their own experiences and information so that other members learn from each other.

**Continuing professional development:** The compulsory or voluntary, formal or informal process by which individuals update their professional competences to develop the personal qualities required in their professional lives.

**Cooperation:** A situation in which individuals interact together in order to accomplish a specific goal or develop an end product.

**CSCL:** A disciplinary area which focuses on how both technology and collaborative learning can facilitate knowledge-sharing and lead to enhanced peer interaction.

**CSCW:** An interdisciplinary field that helps people design, implement, and use technical systems that support working cooperatively; CSCL is referred to a subfield of CSCW throughout the literature.

**Distance learning:** Education that occurs via means of electronic media linking instructors and students who are located in different settings.

**Digital learning:** Any digital learning resource or activity that are made accessible through the means of the Internet. Materials put onto the corporate learning environment along with the use of weblogs, forums, wiki pages, e-mail, audio files such as podcasts, or online libraries can be cited as popular examples of digital learning.

**E-learning:** Learning conducted via use of digital media, typically on the Internet.

**E-mail:** Messages exchanged via electronic means across dispersed computer networks.

**Facebook:** One of the most popular free social networking websites that provides its users with the opportunity to create profiles, upload photos and video, send messages and interact with their 'friends'.

**Formal learning:** Learning occurring within an organization or other context with the purpose of providing education or training. A pre-defined curricula, different roles of teacher and learner are among its distinguishing features.

**Groupware:** Software designed to support individuals involved in a shared task or goal.

**Informal learning:** Learning that takes place via means of books, self-study programs, online resources, or communities of practice.

**Information literacy:** The skills required to be able to identify, locate, assess, and make efficient use of relevant information for the issue or problem at hand.

**Instant messaging (IM):** A form of communications service that provides one with the ability to communicate with others in real time over the Internet by making use of written rather than oral communication.

**Internet:** A global network that consists of millions of computers connected to each other.

**Knowledge:** The understanding we have of the world—the individual and collective "know-how" and truth constructed and reconstructed based on interaction among others.

**Learning:** The process of building new knowledge and/or acquiring understanding via such means as experience, new combinations of prior knowledge, interaction and reflection, dialogue and discussion, observation, and practice.

**Learning management system:** A software application used to register, report, keep track of and deliver e-learning courses.

**Lotus Connect:** An online platform that consists of a corporate directory tool, a social bookmarking service and a blog aggregation service.

**Lotus Notes:** IBM's Messaging and groupware software including tools for mailing, sharing files, collaborative discussions, keeping agendas, and scheduling tasks.

**Lotus Sametime:** A tool that supports enterprise software and business process integration.

**Non-formal learning:** Type of learning for which there is no predefined curriculum or syllabus which are usually related to 'formal learning', yet are better structured than 'informal learning'.

**Online collaboration:** Collaboration by use of digital tools such as discussion forums, weblogs, wiki pages, or online communities utilized by various individuals to reach a common goal.

**Organizational learning:** An enterprise-wide ongoing process that improves its ability to acknowledge, sense, and probe to both internal and external change.

**Podcast:** A digital medium consisting of audio, video, or PDF, files subscribed to and downloaded through RSS feeds.

**Real Simple Syndication (RSS):** A web feed format that provides the opportunity to publish frequently updated information: blog entries or news or audio, video.

**Skype:** An Internet telephony service that also allows instant messaging in real time.

**Social network:** A dedicated website or other application which provides the users with the opportunity to communicate with each other by posting information, comments, messages, images, etc.

**Social software:** Social communication tools which are based on the Internet technology and which aim to enhance communication and collaboration.

**Social web:** A set of social relations providing links for users of the World Wide Web. Its main fields of interest are the design and development of web pages and applications are developed with the aim of improving social interactions.

**Universal Serial Bus (USB):** A device that supports data transfer rates of 12 Mbp.

**Virtual private network (VPN):** A network relying on the Internet and public telecommunication infrastructure to enable remote individual users to gain access to their organizational network.

**Virtual learning environments (VLE):** A web-based learning and teaching system which provides online access to classes, curriculum, exams, assignments, and other grades.

**Web X:** A generic term which denotes paradigm shifts from the perspectives of both the utilization of the world wide web and underlying technological changes.

**Web 1:** The term used to refer to the primary stage of the world wide web which allowed the web users only to seek information and read it ("read-only web").

**Web 2.0:** The term used to refer to the secondary stage of the world wide web which provided the users with the opportunity to contribute content and interact with others ("read-write web").

**World Wide Web:** The whole set of documents existing on all Internet servers which make use of the HTTP protocol and which can be accessed by users through a simple point-and-click system.

**Wiki:** A Web site that allows users to make corrections or contributions.

**Workplace learning:** Traditional classroom training or online training that people receive when they are at work.

# References

[1] Abbitt, J.T. and Klett, M.D. (2007) Identifying influences on attitudes and self-efficacy beliefs towards technology integration among pre-service educators. *Electronic Journal for the Integration of Technology in Education*, 6, 28–42 (Online). Available at: http://ejite.isu.edu/Volume6/Abbitt.pdf (Accessed 15 September 2010).

[2] Alexander, B. (2006) Web 2.0: A new wave of innovation for teaching and learning? *EDUCAUSE Review*, 41(2), 33–44.

[3] Anderson, P. (2007) *What is Web 2.0?* Ideas, technologies and implications for education. *JISC* (Online). Available at: www.jisc.ac.uk/media/documents/techwatch/tsw0701b.pdf (Accessed 10 June 2010).

[4] Anderson, K., Dittlau, A., and Funkel, K. (2004) Probing students. In M. Agger Eriksen, L. Malmborg, J. Nielsen (eds.) *CADE2004 Web Proceedings of Computers in Art and Design Education Conference*. Copenhagen Business School, Denmark and Malmö University, Sweden, 29 June–1 July 2004. Copenhagen: CADE.

[5] Arnold, R. and Schussler, I. (1998) *Wandel der Lernkulturen: Ideen und Bausteine für Ein Lebendiges Lernen*. Darmstadt: Wissenschaftliche Buchgesellschaft.

[6] Bannon, L., Bjorn-Andersen, N., and Due-Thomsen, B. (1988) Computer support for cooperative work: An Appraisal and Critique. In *Proceedings EURINFO 88 – Information Technology for Organizational Systems*. Amsterdam: North-Holland, pp. 297–303.

[7] Bannon, L. and Schmidt, K. (1991) CSCW: Four characters in search of a context. In J. Bowers and S. Benford (eds.) *Studies in Computer Supported Cooperative Work: Theory, Practice and Design*. Amsterdam: North-Holland, pp. 3–16.

[8] Barton, L. (2005) Emancipatory research and disabled people: some observations and questions. *Educational Review*, 57(3), 317–327.

[9] Baylor, A. L., Shen, E., and Warren, D. (2004) Supporting learners with math anxiety: The impact of pedagogical agent emotional and motivational support. *Paper presented at the Workshop on "Social*

*and Emotional Intelligence in Learning Environments". International Conference on Intelligent Tutoring Systems* (Online). Maceió, Brazil. Available at: http://ritl.fsu.edu/papers/W7_baylor_revised.pdf (Accessed 1 October 2010).

[10] Bica, F., Verdin, R. and Maria-Vicari, R. (2005) Inteli Web: Adaptation of the self-efficacy in an intelligent e-learning systems. In *Proceedings of the Fifth IEEE International Conference on Advanced Learning Technologies* (Online). IEEE. Available at: http://ieeexplore.ieee.org/iel5/10084/32317/01508741.pdf?tp=&isnumber=&arnumber=1508741 (Accessed 20 October 2010).

[11] Billet, S. (2001) *Learning in the Workplace: Strategies for effective practice.* Crows Nest, NSW: Allen and Utwin.

[12] Bock, G. and Marca, D. (1995) *Designing Groupware.* New York: McGraw-Hill.

[13] Boud, D. and Solomon, N. (2001) *Work-based Learning: A New Higher Education?* Buckingham: Open University Press.

[14] Boyd, D. (2007). The significance of social software. In T. N. Burg and J. Schmidt (eds.), *Blog Talks reloaded: Social Software Research and Cases.* Norderstedt, Germany: Books on Demand, pp. 15–30.

[15] Brown, J.S. (1993) Thinking, working and learning. In M. Crossan H.W. Lane and R.E. White (eds.) *Learning in Organizations.* London, ONT: Western Business School Monograph, pp. 81–99.

[16] Brown, J.S. and Duguid, P. (1991) Organizational learning and communities-of-practice: Toward a unified view of working, learning, and innovating. *Organization Science*, 2(1), 40–57.

[17] Brown, J.S. and Duguid, P. (1998) Organizing knowledge. *California Management Review*, 40(3), 90–111.

[18] Brown, J.S. and Duguid, P. (2000) *The Social Life of Information.* Boston: Harvard Business School Press.

[19] Caballé, S. (2011) Supporting collaborative learning discussions on asynchronous time: A technological perspective. *eLC Research Paper Series*, 2, 45–57.

[20] Cedefop (2003) *The challenge of e-learning in small enterprises: Issues for policy and practice in Europe.* Brussels: Cedefop Panorama series.

[21] Chi-Sum, W., Law, K.S. and Guo-hua, H. (2008) On the importance of conducting construct-level analysis for multidimensional constructs in theory development and testing. Journal of Management, 34(4), 744–764.

[22] Ciborra, C.U. (ed.) (1996) *Groupware & Teamwork: Invisible Aid or Technical Hindrance*. Chicester: John Wiley & Sons.
[23] Clark, H.H. (1996). *Using Language*. Cambridge: Cambridge University Press.
[24] Cogburn, D.L. (2002) HCI in the so-called developing world: What's in it for everyone. *ACM Interactions*, 10(2), 80–87.
[25] Cohen, W.M. and Levinthal, D.A. (1990) Absorptive capacity: A new perspective on learning and innovation. *Administrative Science Quarterly*, 35(1), 128–152.
[26] Cole, M. and Griffin, P. (Eds.) (1987) *Contextual Factors in Education*. Madison: Wisconsin Center for Educational Research.
[27] Collins, A. (1984) Cognition and computers attack education. Invited address, Division 15, APA Annual Meeting, Toronto.
[28] Collis, B. and Moonen, J. (2001) *Flexible Learning in A Digital World: Experiences and Expectations*. London: Kogan Page.
[29] Conole, G. (2007) Describing learning activities: tools and resources to guide practice. In H. Beetham and R. Sharpe (eds.) *Rethinking Pedagogy for a Digital Age: Designing and Delivering E-learning*. London: Routledge, pp. 81–91.
[30] Conole, G. and Fill, K. (2005) *The learning taxonomy in Dialogus Plus Toolkit* (Online). University of South Hampton. Available at: http://joker.ecs.soton.ac.uk/dialogplustoolkit/userarea/ (Accessed 28 May 2010)
[31] Cook, S.D.N. and Brown, J.S. (1999) Bridging epistemologies: The generative dance between organizational knowledge and organizational knowing. *Organization Science*, 10(4), pp. 381–400.
[32] Cornwall, A. and Jewkes, R. (1995) What is participatory research? *Social Science and Medicine*, 41(12), 1667–1676.
[33] Creanor, L., Trinder, K., Gowan, D. and Howells, C. (2006) LEX: The learner experience of e-learning – final project report, August 2006, JISC.
[34] Crossan, M.M., Lane, H.W. and White, R.E. (1999) An organizational learning framework: From Intuition to Institution. *Academy of Management Review*, 24(3), 522–537.
[35] Daft, R. and Huber, G.P. (1987) How organizations learn: A communications framework. *Research in the Sociology of Organizations*, 5, 1–36.

[36] Daft, R. and Weick, K.E. (1966) Toward a model of organizations as interpretation systems. *Academy of Management Journal*, 9(2), 284–295.

[37] Darr, E.D. and Goodman, P.S. (1995) New research opportunities in computer-aided learning systems. In D.R.C. Cooper (ed.) *Trends in Organizational Behavior*. New York: Wiley, pp. 90–101.

[38] Davenport, E. and Hall, H. (2002) Organizational Knowledge and Communities of Practice. *Annual Review of Information Science and Technology (ARIST)*, 36, pp. 171–227.

[39] Davis, F. (1989) Perceived usefulness, perceived ease of use, and user acceptance of information technology. *MIS Quarterly*, 13(3), 319–339.

[40] Davies, R., Marcella, S., McGrenere, J., and Purves, B. (2004) The ethnographically informed participatory design of a PDA application to support communication. In *Proceedings of ACM ASSETS 2004*, 153–160 (Online). Available at: http://www.cs.ubc.ca/~joanna/papers/ASSETS2004_Davies.pdf (Accessed 21 April 2010).

[41] Dennis, A., Venkatesh, V. and Ranesh, V. (2003) *Adoption of Collaboration Technologies: Integrating Technology Acceptance and Collaboration*. Bloomington: Rob Kling Center for Social Informatics Speaker Series.

[42] Denzin, N.K. (1970) *The Research Act in Sociology: A Theoretical Introduction to Sociological Methods*. London: Butterworths.

[43] Devlin, T. (1993) Distance Training. In D. Keegan (ed.) *Theoretical Principles of Distance Education*. London: Routledge Falmer, pp. 60–74.

[44] Dewey, J., (1938) *Logic: The Theory of Inquiry*. New York: Holt.

[45] Dewhurst, F.W. and Navarro, C. (2004) External communities of practice and relational capital. *The Learning Organization*, 11(4), 322–331.

[46] Dewsbury G., Clarke K., Rouncefield M., and Sommerville I. (2004) Depending on digital design: extending inclusivity. *Housing Studies*, 19(5), 811–825.

[47] Dillenbourg, P. (1999) What do you mean by "collaborative learning"? In P. Dillenbourg (ed.) *Collaborative Learning: Cognitive and Computational Approaches*. Amsterdam: Pergamon, pp. 1–16.

[48] Dirckinck-Holmfeld, L. (2002) Designing virtual learning environments based on problem oriented project pedagogy in L. Dirckinck-Holmfeld and B. Fibiger (Eds.), *Learning in virtual environments*. Fredriksberg Denmark, Samfundslitteratur, pp. 31–54.

[49] Dohn, N. (2010) Web 2.0: Inherent tensions and evident challenges for education. *International Journal of Computer-Supported Collaborative Learning*, 4(3), 343–363.

[50] Downes, S. (2010) *E-learning 2.0* (Online). National Research Council of Canada. Available at: http://www.elearnmag.org/subpage.cfm?section=articles&article=29-1 (Accessed 20 May 2010).

[51] Druin, A. (2007) *Connecting Generations: Developing Co-Design Methods for Older Adults and Children* (Online). Available at: http://hcil.cs.umd.edu/trs/2007-15/2007-15.pdf (Accessed 16 December 2009).

[52] Duckett, P. S., and Fryer, D. (1998) Developing empowering research practices with people who have learning disabilities. *Journal of Community and Applied Social Psychology* (8), 57–60.

[53] Duckett, P. and Pratt, R. (2007) The emancipation of visually impaired people in social science research practice, *British Journal of Visual Impairment*, 25(1), 5–20.

[54] Duncan, R.B. and Weiss, A. (1979) Organizational learning: Implications for organizational design. In B. Straw (ed.) *Research in Organizational Behavior*. Greenwich: JAI Press, pp. 75–124.

[55] Dyke, M., Conole, G., Ravenscroft, A., and de Freitas, S. (2007) Learning theories and their application to e-learning. In G. Conole and M. Oliver (eds.) *Contemporary Perspectives in E-learning Research: Themes, Methods and Impact on Practice*. London: Routledge Falmer, pp. 63–80.

[56] Easterby-Smith, M. and Lyles, M.A. (2003) *The Blackwell Handbook of Organizational Learning and Knowledge Management*. Oxford: Blackwell.

[57] Edwards, R., and Usher, R. (2008) *Globalisation and pedagogy: Space, place and identity* (2nd ed.). Milton Park, England: Routledge.

[58] Elkjaer, B. and Wahlgren, B. (2006) Organizational learning and workplace learning-similarities and differences. In P. Jarvis, V. Andersen, B. Elkjaer, S. Hoyrup, and E. Antonacopoulou (eds.) *Learning, Working and Living – Mapping the Terrain of Working Life Learning*. New York: Palgrave Macmillan, pp. 87–101.

[59] Ellis, C.A., Gibbs, S.J. and Rein, G.I. (1992) Groupware: Some issues and experiences. In G.B. D. Marca (ed.) *Groupware: Software for Computer Supported Cooperative Work: Assisting Human-human Collaboration*. Los Alamitos: IEEE Computer Society Press.

[60] Ely, D.P. (1990) Conditions that facilitate the implementation of educational technology innovations. *Journal of Research on Computing in Education*, 23, 298–305.
[61] Engeström, Y. (1999) Activity theory and individual and social transformation. In Y. Engeström, R. Miettinen and R. L. Punamäki (eds.), *Perspectives on activity theory*. Cambridge: Cambridge University Press, pp. 19–38.
[62] Engestrom, Y. (2001) Expansive learning at work: Toward an activity theoretical reconceptualization. *Journal of Education and Work*, 14, 133–156.
[63] Engestrom, Y. and Middleton, D. (1996) Introduction: Studying work as mindful practice. In Y. Engestrom and D. Middleton (eds.), *Cognition and Communication at Work*. Cambridge: Cambridge University Press, pp. 1–15.
[64] Etzioni, A. (1975) An Engineer-Social Science Team at Work. *Technology Review*, 1975, 26–31.
[65] Everitt, A., Hardiker, P., Littlewood, J. and Mullender, A. (1992) *Applied Research for Better Practice*, London: Macmillan.
[66] Facebook. (2014) Statistics (Online). Retrieved at http://www.facebook.com/press/info.php?statistics. (Accessed 15 January 2014).
[67] Fajerman, L. and Treseder, P. (2000) *Children are Service Users too: A Guide to Consulting Children and Young People*. London: Save The Children.
[68] Finholt, T. and Sproull, L.S. (1990) Electronic groups at work. *Organization Science*, 1(1), 41–64.
[69] Fiol, C. M. and Lyles, M. A. (1985) Organizational learning. *Academy of Management Review*, 10(4), 803–813.
[70] Fischer, G., and Ostwald, J. (2002) Seeding, evolutionary growth, and reseeding: Enriching participatory design with informed participation. In *Proceedings of the Participatory Design Conference (PDC'02)*, Malmo, Sweden, pp. 135–143.
[71] Flate, M., Rubia, B., Suárez, C., Román, P., Area, M., Garcia, I. and Domingo, J. (2011). Some approaches to the effects of time on collaborative online learning. *eLC Research Paper Series*, 2, pp. 73–82.
[72] Foth, M. (2004) Animating personalised networking in a student apartment complex through participatory design. In *Proceedings of the Participatory Design Conference*, Toronto (Online). Available at: http://eprints.qut.edu.au/archive/00001904/01/foth_id172_shortpaper.pdf (Accessed 16 June 2010).

[73] French, S. and Swain, J. (2004) Researching together: A participatory approach. In S. French and J. Sim, (eds.), *Physiotherapy: A Psychosocial Approach*, 3rd edition, Oxford: Butterworth–Heinemann, pp. 50–64.

[74] Gee, J.P. (2004) *Situated Language and Learning: A Critique of Traditional Schooling*. New York: Palmgrave-McMillan.

[75] Gibb, T., and Fenwick, T. (2008) 'Older' professionals' learning in an ageistculture: Beneath and between the borders. In J. Groen and S. Guo (Eds.), *Online Proceedings of the Canadian Association for the Study of Adult Education*. Vancouver, British Columbia (Online). Available at: <http://www.oise.utoronto.ca/CASAE/cnf2008/OnlineProceedings-20 08/CAS2008-Gibb.pdf> (Accessed 25 April 2010).

[76] Giddens, A. (1984). *The constitution of society*. Berkeley: University of California Press.

[77] Gillman, M., Swain, J. and Heyman, B. (1997) 'Life' history or 'Case' history: the objectification of people with learning difficulties through the tyranny of professional discourses. *Disability* and *Society*, 12(5), 675–694.

[78] Golensky, M. (Ed.). (2000) Volunteerism. ARNOVA Abstracts, 23 (4). Indianapolis, IN: Association for Research on Nonprofit Organizations and Voluntary Action.

[79] Goodyear, P. and Zenios, M. (2007) Discussion, collaborative knowledge work & epistemic fluency. *British Journal of Educational Studies*. 55(4), pp. 351–368.

[80] Green, J. (1999) Qualitative Methods. *Community Eye Health*, 12(31), 46–47.

[81] Hanson, E., Magnusson, L., Arvidson, H., Claeson, A., Keedy, J. and Nolan, M. (2007) Working together with persons with early stage dementia and their family members to design a user-friendly technology based support service. *Dementia*, 6(3), 411–434.

[82] Hartson, H. R. (2003) Cognitive, physical, sensory and functional affordances in interaction design. *Behaviour and Information Technology*, 22(5), 315–338.

[83] Haythornthwaite, C. (2008) Ubiquitous transformations. Proceedings of the 6th International Conference on Networked Learning (Online; pp. 598–605). Available at: http://www.networkedlearningconference.o rg.uk/past/nlc2008/Info/confpapers.htm#Top (Accessed 10 January 2011).

[84] Hazemi, R., Hailes, S., and Wilbur, S. (eds.) (1998) *The Digital University: Reinventing the Academy*. London: Springer.

[85] Hedberg, B. (1981) How organizations learn and unlearn. In P.C. Nystrom and P.H. Starbuck (eds.) *Handbook of Organizational Design*. New York: Oxford University, pp. 3–27.

[86] Hegel, G. W. F. (1967) Phenomenology of spirit (J. B. Baillie, Trans.). New York, NY: Harper and Row. (Original work published in 1807).

[87] Henri, F. and Pudelko, B. (2003) Understanding and analyzing activity and learning in virtual communities. *Journal of Computer Assisted Learning*, 19(4), 474–487.

[88] Hess, D., Rogovsky, N., and Dunfee, T. (2002) The next wave of corporate community involvement. *California Management Review*, 44(3), pp. 110–125.

[89] Hicks, D. (1996). Contextual inquiries: A discourse-oriented study of classroom learning. In D. Hicks (ed.), *Discourse, learning and schooling* (pp. 104–141). New York: Cambridge University Press.

[90] Hiltz, R. (1990) Evaluating the virtual classroom. In online education: perspectives on a new environment. In Harasim L.M. and Turnoff M. (eds.). *Online Education: Perspectives On A New Environment*, New York: Praeger, pp. 133–183.

[91] Hiltz, S.R. and Johnson, K. (1989) Measuring acceptance of computer mediated communication systems. *Journal of the American Society for Information Science*, 40(6), 386–397.

[92] Hiltz, S.R. and Turoff, M. (1978) *The Network Nation: Human Communication via Computer*. 1st ed. Cambridge, MA: MIT Press.

[93] Hopkins, J. (2011) Assessed real-time language learning tasks online: how do learners prepare? eLC Research Paper Series, 2, pp. 59–72.

[94] Houle, C.O. (1972) *The Design of Education*. San Francisco: Jossey-Bass.

[95] Huber, G.P. (1991) Organizational learning: The contributing processes and the literatures. *Organization Science*, 2(1), 88–115.

[96] Hutchins, E (1991) The social organization of distributed cognition. In L. B. Resnick, J. M. Levine and S. D. Teasley (eds.), *Perspectives on Socially Shared Cognition*. (pp. 283–307) Washington DC, USA: American Psychological Association.

[97] Illeris, K., Andersen, V., Bottrup, P., Clematide, B., Dirckinckholmfeld, L., Elkjaer, B., Elsborg, S., Hoyrup, S., Jorgensen, C.H., Kanstrup, N. and Aarkrog, V. (2004) *Learning in Working Life*. Roskilde: Roskilde University Press.

[98] Johansen, R. (1988) Groupware: Computer Support for Business Teams. New York: The Free Press.

[99] Jones, N.B. and Laffey, J. (2002) How to facilitate e-collaboration and e-learning in organizations. A. Rosset (ed.) In *The ASTD E-Learning Handbook*. New York, USA: McGraw-Hill, pp. 80–101.

[100] Karahanna, E., Straub, D., and Chervany, N. (1999) Information technology adoption across time: A cross-sectional comparison of pre-adoption and post-adoption beliefs. *MIS Quarterly*, 23(2), 183–213.

[101] Kasworm, C. (1997) *The Agony and The Ecstasy of Adult Learning: Faculty Learning Computer Technology*. Cincinnati, OH: The American Association for Adult and Continuing Education.

[102] Kazmer, M.M. (2007) Community-embedded learning. In R. Andrews and C. Haythornthwaite (Eds.), *Handbook of E-learning Research* (pp. 311–327). London: Sage.

[103] Kiesler, S., Siegel, J., and McGuire, T. (1984) Social psychological aspects of computer-mediated communication. *American Psychologist* 39, 1123–1134.

[104] Kilpatrick, R., McCartan, C., McAlister, S. and McKeown, P. (2007) 'If I am brutally honest, research has never appealed to me ...' The Problems and Successes of a Peer Research Project, Educational Action Research, 15(3), pp. 351–369.

[105] Kuhn, T. (1962) *The Structure of Scientific Revolutions*. Chicago, US: University of Chicago Publication.

[106] Kurland, D.M. and Kurland, L.C. (1987) Computer Applications in Education: A Historical Overview. In *Annual Review of Computing Science*, Vol. 2, pp. 317–358. Palo Alto, CA: Annual Reviews Inc.

[107] Lathlean, J. and Le May, A. (2002) Communities of Practice: An opportunity for Interagency Working. *Journal of Clinical Nursing*, 11(3), 394.

[108] Lave, J. (1988) *Cognition in Practice: Mind, mathematics and culture in everyday life*. New York: Cambridge University Press.

[109] Lave, J. and Wenger, E. (1991) *Situated Learning: Legitimate Peripheral Participation*. Cambridge: Cambridge University Press.

[110] Lave, J. (1996) Teaching, as learning, in practice. *Mind, Culture, and Activity*, 3(3), 149–164.

[111] LeBaron, C. (2002) Technology does not exist independent of its use. In T. Koschmann, R. Hall and N. Miyake (eds.), *CSCL 2: Carrying forward the conversation*. Mahwah, NJ: Lawrence Erlbaum Associates, pp. 433–439.

[112] Leontiev, A. N. (1978) *Activity, consciousness, and personality*. Englewood Cliffs, NJ: Prentice-Hall.

[113] Lesser, E. and Prusak, L. (2000) Communities of practice, social capital and organizational knowledge. In E. L. Lesser, M. A. Fontaine, and J. A. Slusher (eds.) *Knowledge and Communities*. Boston: Butterworth Heinemann, pp. 123–132.

[114] Levitt, B. and March, J. G. (1988) Organizational learning. *Annual Review of Sociology*, 14(1), 319–338.

[115] Lewin, K. (1952) Group decision and social change. In T.M. Newcomb and E.L. Hartley (eds.) *Readings in Social Psychology*. New York: Henry Holt and Company, pp. 83–95.

[116] Lou, H. (1995) Groupware at work: users' experience with Lotus Notes. *Journal of End User Computing*, 6(3), 13–19.

[117] Luria, A. R. (1976) *Cognitive Development: Its Cultural and Social Foundations*. Cambridge: Harvard University Press.

[118] Malcolm, J., Hodkinson, P. and Colley, H. (2003) The interrelationships between informal and formal learning. *Journal of Workplace Learning*, 15(7/8), 313–318.

[119] Malhotra, A. and Majchrzak, A. (2005) Virtual workspace technologies. *MIT Sloan Management Review*, 46(2), 11–14.

[120] Mandl, H. and Krause, U.M. (2001) *Lernkompetenz für die Wissensgesellschaft*. München: Forschungsbericht Nr. 145.

[121] Marx, K. (1976) Capital (B. Fowkes, Trans. Vol. I). New York, NY: Vintage.

[122] Mason, J. (2002) *Qualitative Researching*, 2nd ed. London: Sage.

[123] Mayes, T. (2006) LEX Methodology Report (Online). *JISC* Available at: http://www.jisc.ac.uk/media/documents/programmes/elearningpedagogy/lex_method_final.pdf. (Accessed 20 June 2010).

[124] Mclellan, H. (1997) Creating virtual communities on the Web. In Khan, B.H. (ed.) *Web-based instruction: Development, Application, and Evaluation*. Englewood Cliffs, NJ: Educational Technology Publications, pp. 185–190.

[125] Munford, R., Sanders, J., Veitch, B.M. and Conder, J. (2008) Looking inside the bag of tools: Creating research encounters with parents with an intellectual disability. *Disability* and *Society*, 23(4), 337–347.

[126] Myers, M. (1991) Cooperative learning in heterogeneous classes. *Cooperative Learning*, 11(4).

[127] Naylor, P., Wharf-Higgins, J., Blair, L., Green, L. and O'Connor, B. (2002) Evaluating the participatory process in a community based heart health project. *Social Science and Medicine*, 55, 1173–1187.

[128] Neilson, R. (1997) *Collaborative Technologies* and *Organizational Learning*. Hershey, PA: Idea Group Publishing.
[129] Nelson, G., Ochocka, J., Griffin, K. and Lord, J. (1998) Nothing about me, without me: Participatory Action Research with self-help/mutual aid organizations for psychiatric consumer/survivors. *American Journal of Community Psychology*, 26(6), 881–912.
[130] Newell, A., Carmichael, J. and Morgan, M. (2007) Methodologies for involving older adults in the design process. In *Proceedings of the 4th International Conference on Universal Access in HCI* (Online). Available at: http://www.springerlink.com/content/53t5026735v65721/fullte xt.pdf (Accessed 21 April 2010).
[131] Newman, D., Griffin, P. and Cole, M. (1989) *The Construction Zone: Working for Cognitive Change in School*. New York: Cambridge University Press.
[132] Nightingale, C. (2006) *Nothing About Me, Without Me. Involving Learners with Learning Difficulties or Disabilities*. London: Learning Skills Development Agency.
[133] Norman, D. A. (1988) *The Psychology of Everyday Things*. New York: Basic Books.
[134] O'Donnell, D., Porter, G., McGuire, D., Garavan, T.N., Heffernan, M. and Cleary, P. (2003) Creating intellectual capital: A Habermasian Community of Practice (CoP). *Journal of European Industrial Training*, 27(2/3/4), 80–87.
[135] Olson, G.M., Teasley, S., Bietz, M.J. and Cogburn, D.L. (2002) Collaboratories to support distributed science: The example of international HIV/AIDS research. *Proceedings of the 2002 annual research conference of the South African institute of computer scientists and information technologists on enablement through technology*, Cape Town: University of South Africa, pp. 44–48.
[136] O'Reilly, T. (2005) *What is Web 2.0: design patterns and business models for the next generation of software* [Online]. Available at: <http://www.oreillynet.com/pub/a/oreilly/tim/news/2005/09/30/what-is-web-20.html> (Accessed 30 January 2010).
[137] O'Reilly, T. (2007) *Web 2.0 Compact Definition: Trying Again* (Online). Available at: http://www.oreillynet.com/pub/a/oreilly/tim/news/2005/0 9/30/what-is-web-20.html (Accessed 30 January 2010).
[138] Orlikowski, W. (1992) The duality of technology: Rethinking the concept of technology in organizations. *Organization Science*, 3(3), 398–427.

[139] Oosteveen, A. and van de Besselaar, P. (2004) From small scale to large scale user participation: A case study of participatory design in e-government systems. In Proceedings of Participatory Design Conference 2004, Toronto, Canada.

[140] Pain, R. and Francis, P. (2003) Reflections on participatory research. *Area*, 35(1), 46–54.

[141] Patton, M.Q. (1980) *Qualitative Evaluation Methods*. Beverly Hills, CA, USA: Sage Publications.

[142] Patton, M.Q. (1990) *Qualitative Evaluation and Research Methods*. Newbury Park; London, UK: Sage.

[143] Pea, R. D. (1994). Seeing what we build together: Distributed multimedia learning environments for transformative communications. *The Journal of the Learning Sciences*, 13(3), pp. 285–299.

[144] Pérez-Mateo M. and Guitert, M. (2011) Social aspects as regards the time factor: An analysis of the work process in a virtual group. *eLC Research Paper Series*, 2, pp. 29–44.

[145] Piaget, J. (1966) *Psychology of Intelligence*. Totowa, NJ: Littlefield, Adam and Co.

[146] Polanyi, M.F.D. (2002) Communicative Action in Practice: Future Search and The Pursuit of An Open, Critical and Non-coercive Large-group Process. *Systems Research* and *Behavioural Science*, 19(4), pp. 357–366.

[147] Putnam, R. D. (2000) *Bowling alone: The collapse and revival of American community*. New York: Simon and Schuster.

[148] Radermacher, H. (2006) *Participatory Action Research with People with Disabilities: Exploring Experiences Of Participation*. Melbourne: Victoria University.

[149] Reason, P. and Heron, J. (1986) Research with people: The paradigm of co-operation experiential enquiry. *Person-Centred Review*, 1(4), 456–476.

[150] Reiser, B. (2002) Why scaffolding should sometimes make tasks more difficult for learners. In G. Stahl (Ed.), *Computer support for collaborative learning: Foundations for a CSCL community*. Proceedings of CSCL 2002. Hillsdale, NJ: Erlbaum.

[151] Richardson, M. (2000) How we live: participatory research with six people with learning difficulties. *Journal of Advanced Nursing* 32(6), 1383–1395.

[152] Riddle, M. and Arnold, M. (2007) The day experience resource kit (Online). Available at: http://www.matthewriddle.com/papers/Day_Experience_Resource_Kit.pdf (Accessed 30 January 2010).
[153] Rocco, E. (1998) Trust breaks down in electronic contexts but can be repaired by some initial face-to-face contact. In *Conference Proceedings on Human Factors in Computing Systems*, Los Angeles, CA, April, 1998, ACM Press, New York, pp. 496–502.
[154] Rockwood, R. (1995) Cooperative and collaborative learning. *National Teaching and Learning Forum*, 4(6), 8–17.
[155] Rogers, E. (1962) *Diffusion of Innovations* New York: Free Press.
[156] Rogoff, B. (1990) *Apprenticeship in Thinking: Cognitive Development in Social Context*. Oxford: Oxford University Press.
[157] Romero, M. (2011) The time factor in an online group course from the point of view of its students. *eLC Research Paper Series*, 2, 17–28.
[158] Roschelle, J. and Teasley, S. (1995) The construction of shared knowledge in collaborative problem solving. In C. O'Malley (Ed.), *Computer-supported collaborative learning*. Berlin, Germany: Springer Verlag, pp. 60–197.
[159] Rosenberg, M.C. (2006) *Beyond E-Learning*. San Francisco: Pfeiffer.
[160] Rosenberg, M.C. (2001) *E-learning – Strategies for Delivering Knowledge in the Digital Age*. New York: McGraw-Hill.
[161] Ryberg, T. (2008) Challenges and potentials for institutional and technological infrastructures in adopting social media. In *Proceedings of the 6th International Conference on Networked Learning* (pp. 658–665; online). Available at: http://www.networkedlearningconference.org.uk/past/nlc2008/Info/confpapers.htm#Top (Accessed 08 August 2010).
[162] Ryberg, T. and Christiansen, E. (2008) Community and social network sites as technology enhanced learning environments. *Technology, Pedagogy and Education*, 17, 207–219.
[163] Ryberg, T., Glud, L.N., Buus, L. and Georgsen, M. (2010) Identifying Differences in Understandings of PBL, Theory and Interactional Interdependencies. *Networked Learning Conference 2010*, Aalborg.
[164] Ryberg, T., and Ponti, M. (2005) Constructing Place: The relationship between place-making and sociability in networked environments – a condition for productive learning environments. In L. Dirckinck-Holmfeld, B. Lindström, B. M. Svendsen and M. Ponti (eds.), *Conditions for Productive Learning in Networked Learning Environments*. Aalborg University/Kaleidoscope, pp. 90–98.

[165] Salomon, G. (1993) *Distributed Cognitions—Psychological and Educational Considerations.* Cambridge: Cambridge University Press.
[166] Säljö, R. (1995). Mental and Physical Artifacts in Cognitive Practices. In P. Reimannand H. Spada (eds.). *Learning in Humans and Machines: Towards An Interdisciplinary Learning Science.* Oxford, Elsevier Science, pp. 90–99.
[167] Samarawickrema, G. (2007) Piloting social networking and Web 2.0 software at Deakin University. In *ICT: Providing choices for learners and learning. Proceedings Ascilite Singapore 2007* (Online). Available at: http://www.ascilite.org.au/conferences/singapore07/procs/samarawickrema.pdf (Accessed 28 August 2010).
[168] Scardamalia, M., and Bereiter, C. (1996) Computer support for knowledge-building communities. In T. Koschmann (ed.), *CSCL: Theory and practice of an emerging paradigm.* Hillsdale, NJ: Lawrence Erlbaum, pp. 249–268.
[169] Schrage, M. (1990) *Shared Minds: The New Technologies of Collaboration.* New York: Random House.
[170] Seale, J., Draffan, E.A and Wald, M. (2008) *An evaluation of the use of participatory methods in exploring disabled learners' experiences of e-learning.* LEXDIS Methodology Report to JISC (Online). Available at: http://www.lexdis.org/project/media/LEXDIS%20Methodology%20Report%20Seale%20Draffan%20Wald%20July%2008.doc (Accessed 25 August 2010).
[171] Selwyn, N. (2006) Digital division or digital decision? A study of non-users and low-users of computers, poetics. *Journal of Empirical Research in Culture, Media and the Arts*, 34(4–5), 273–292.
[172] Sfard, A. (1998) On two metaphors for learning and the dangers of choosing just one. *Educational Researcher*, 27 (2), pp. 4–13.
[173] Sharpe, R., Benfield, G., Lessner, E., and De Cicco, E. (2005) *Learner Scoping Study: Final Report* [Online]. JISC. Available at: http://www.jisc.ac.uk/uploaded_documents/scoping%20study%20final%20report%20v4.1.doc (Accessed 21 April 2010).
[174] Shrivastava, P. (1983) A typology of organizational learning systems. *Journal of Management Studies*, 20 (1), 7–28.
[175] Silva and Breuleux (1994) The Use of Participatory Design in the Implementation of Internet-based Collaborative Learning Activities in K–12 Classrooms (Online). Available at: http://www.helsinki.fi/science/optek/1994/n3/silva.txt (Accessed 29 June 2010).

[176] Sim, J. (1998) Collecting and analysing qualitative data: Issues raised by the focus group. *Journal of Advanced Nursing*, 28(2), 345–352.
[177] Simon, H.A. (1991) Bounded rationality and organizational learning. *Organization Science*, 2(1), 125–134.
[178] Sleeman, D. and Brown, J.S. (Eds.) (1982) *Intelligent Tutoring Systems*. New York: Academic Press.
[179] Smith, J. A. and Osborn, M. (2003) Interpretative Phenomenological Analysis. In A. Smith *Qualitative Psychology*. London: Sage, pp. 40–52.
[180] Soanes, C. and Stevenson, A. (2004) *Concise Oxford English Dictionary*. New York: Oxford University Press.
[181] Spender, J.C. (1996) Organizational knowledge, learning and memory: Three concepts in search of a theory. *Journal of Organizational Change Management*, 9(1), 63.
[182] Spivey, N. (1997) *The Constructivist Metaphor: Reading, Writing, and The Making of Meaning*. San Diego: Academic Press.
[183] Sproull, L. and Kiesler, S. (1995) *Connections: New Ways of Working In the Networked Organization*. Cambridge, MA: The MIT Press.
[184] Stahl, G. (2001) (ed.) *Computer Support for Collaborative Learning: Foundations for a CSCL Community*. Hillsdale, NJ: Lawrence Erlbaum.
[185] Stata, R. (1989) Organizational learning: The key to management innovation. *Sloan Management Review*, 30(3), 63–74.
[186] Steeples, C (2004) Using Action-Oriented or Participatory Research Methods For Research On Networked Learning. In *Paper presented at the Networked Learning Conference 2004*, Lancaster, UK (Online). Available at: http://www.networkedlearningconference.org.uk/past/nlc2 004/proceedings/symposia/symposium4/steeples.htm (Accessed 18 July 2010).
[187] Stewart, R. and Bhagwanjee, A. (1999) Promoting group empowerment and self-reliance through participatory research: A case study of people with physical disability, *Disability* and *Rehabilitation*, 21(7), 338–345.
[188] Suthers, D. (2005) Technology affordances for intersubjective learning: A thematic agenda for CSCL. Paper presented at the international conference of Computer Support for Collaborative Learning (CSCL 2005), Taipei, Taiwan.
[189] Valsiner, J. (1994) Bi-directional cultural transmission and constructive sociogenesis. In W. de Graaf and R. Maier (Eds.) *Sociogenesis re-examined*. New York: Springer, pp. 101–134.

[190] Valsiner J. and van der Veer R. (2000) *The Social Mind: The Construction of An Idea*. Cambridge, UK: Cambridge University Press.
[191] Van Dam, N. (2004) *The E-Learning Fieldbook*. New York: MacGraw-Hill Companies, Inc.
[192] Vygotsky, L. (1978) *Mind in Society: The Development of Higher Psychological Processes*. Cambridge, MA: Harvard University Press.
[193] Waltz, S. B. (2006) Nonhumans unbound: Actor-network theory and the reconsideration of "things" in educational foundations. *Educational Foundations*, 20(3/4), pp. 51–68.
[194] Warner, M. (2004) Building blocks for partnerships. In M. Warner and R. Sullivan (eds.), *Putting partnerships to work* (pp. 24–35). Sheffield, England: Greenleaf Publishing.
[195] Weick, K.E. (1990) Technology as equivoque: Sense-making in new technologies. In P. S. Goodman, L.S. Sproull (eds.) *Technology and Organizations*. San Francisco: Jossey-Bass, pp. 78–93.
[196] Wenger, E. (1999) *Communities of Practice: Learning, Meaning and Identity*. Cambridge: Cambridge University Press.
[197] Wenger, E. (2000) Communities of practice: The key to knowledge strategy. In E. L. Lesser, M. A. Fontaine, and J. A. Slusher (eds.) *Knowledge and Communities*. Boston: Butterworth Heinemann, pp. 3–20.
[198] Wenger, E. (2004) Communities of practice and social learning systems. In K. Starkey, S. Tempest, and A. McKinlay (eds.) *How Organizations Learn: Managing the Search for Knowledge*. London: Thomson, pp. 238–258.
[199] Wenger, E. and Snyder, W.M. (2000) Communities of practice: The organizational frontier. *Harvard Business Review*, 78(1), 139.
[200] Wertsch J.W. (1998) *Mind as Action*. New York: Oxford University Press.
[201] Westera, W. (2004) Implementing integrated e-learning: Lessons learnt from the OUNL case. In W. Jochems Van Merrienboer and R. Koper (eds.) *Integrated E-learning: Implications for pedagogy, technology* and *organization*. London, UK: Routledge Falmer, pp. 56–78.
[202] White, D. and LeCornu, A. (2010) Eventedness and disjuncture in virtual worlds. *Journal of Educational Research*, 52(2), 183–196.
[203] Winograd, T., and Flores C. F. (1986). *Understanding computers and cognition: A new foundation for design*. Norwood, NJ: Ablex.
[204] Young, N. (2006) Distance as a hybrid actor in rural economies. *Journal of Rural Studies*, 22(3), 253–266.

[205] Zack, M.H. and McKenney, J.L. (1995) Social context and interaction in ongoing computer-supported management groups. *Organization Science*, 6(4), 394–422.
[206] Zarb, G. (1992) On the Road to Damascus: First steps towards changing the relations of research production. *Disability, Handicap and Society*, 7(2), 125–38.
[207] Zhao, S. (2001). The increasing presence of teleco presence in the Internet era. In *Paper presented at the annual conference of the American Sociological Association*, Anaheim, CA.

# Index

**C**
computer-supported collaborative learning 15

**E**
e-learning 22, 30, 102

**I**
IBM 2, 12, 33, 57, 102
Informal learning 4, 38, 88, 90

**K**
Knowledge management 11

**O**
Online collaboration 9, 47, 58, 60, 98
Organizational learning 11, 12

**S**
Social learning 15, 16, 31, 85
Social network 7, 64, 73, 84

**W**
Web 2.0 1, 26, 63, 102

# About the Author

Ayse Kok received her MSc in Technology & Learning in University of Oxford in 2006 and her research degree (MLitt) in Technology & Organizational Studies in Oxford in 2015. She participated in various research projects for UN, Nato and the European Commission. Between 2010–2015, Ayse was an adjunct faculty member at Bogazici University in her home town Istanbul. Ayse is also passionate about philanthropy and is the the co-founder of the first non-profit Mooc (http://www.UniversitePlus.com) in Turkey.

Lightning Source UK Ltd.
Milton Keynes UK
UKOW06n1418060416

271689UK00003B/23/P